570
INNO

S0-APN-182

AUG   '00

PLEASE <u>DO NOT</u> REMOVE
CARD FROM POCKET

Innovations in Biology

The sciences move fast—
and ABC-CLIO's Innovations in Science series can help you keep pace.
Each title provides an overview of the events, scientists, and innovations
that have shaped the development of a particular field of science during
the past hundred years. Well suited to student research, or just for
satisfying the curious, these accessible handbooks for the nonspecialist
highlight the scientific breakthroughs of the twentieth century and the
prospects for the twenty-first.

Titles in This Series
Innovations in Astronomy
Innovations in Biology
Innovations in Earth Sciences

INNOVATIONS IN SCIENCE

Innovations in Biology

Overview by Martin Walters

ABC-CLIO

Santa Barbara, California
Denver, Colorado
Oxford, England

Copyright © 1999 by Helicon Publishing Ltd.

All rights reserved. No part of this publication may be reproduced, stored in a retrieval system, or transmitted, in any form or by any means, electronic, mechanical, photocopying, recording, or otherwise, except for the inclusion of brief quotations in a review, without prior permission in writing from the publishers.

Library of Congress Cataloging-in-Publication Data

Innovations in biology.
     p. cm. — (Innovations in science)
  Includes bibliographical references.
   1. Biology.   I. ABC-CLIO II. Series
QH307.2.I56   1999   570—dc21   99-27926

ISBN 1-57607-116-2 (alk. paper)

05 04 03 02 01 00   10 9 8 7 6 5 4 3 2 (cloth)

ABC-CLIO, Inc.
130 Cremona Drive, P.O. Box 1911
Santa Barbara, California 93116-1911

This book is printed on acid-free paper ∞.

Contents

Preface

The study of life—biology—has been a preoccupation of our own species for millennia, yet it is only relatively recently that biology has developed into a powerful science, with implications across a broad range of human activities.

From a rather amorphous subject in its early days, biology blossomed into the rapidly changing, multifaceted subject that we recognize today. In fact, biology now encompasses a wide spectrum of individual disciplines, including, among others, anatomy, animal behavior, biochemistry, biotechnology, cell biology, developmental biology, ecology, evolutionary biology, genetics, paleobiology, physiology, and taxonomy.

The advances in biology during the twentieth century have been remarkable, changing the subject almost out of recognition within a comparatively short period of time. Nineteenth-century biology was essentially based upon natural history, being largely preoccupied with cataloging the diversity of nature—still an important aspect of biology today—but it was far less concerned with the internal processes of living organisms. The emphasis was to shift dramatically during the twentieth century, partly as a result of new apparatus such as the electron microscope, but also as a consequence of the fertile intermingling of ideas and techniques applied in other sciences, notably in organic chemistry.

This application of concepts from other branches of science, such as chemistry, physics, and mathematics, gave biology a much firmer theoretical underpinning than previously, while advances in technology and instrumentation have enabled biologists to investigate much more closely the nature of biological systems, such as the structure and working of individual cells, and even of organelles and other components within cells. In addition, the ever-growing power and sophistication of computers has allowed biologists to analyze their data more quickly and thoroughly, and to simulate the frequently very complex natural systems under investigation.

Innovations in Biology takes a close look at the major innovations in biology that have taken place during the twentieth century. Some of these innovations have come about through the application of techniques and concepts transferred from one discipline to another, while others have resulted more from the focusing of biological research in the light of fresh theoretical insights.

Prominent among such insights was the concept of evolution, developed most notably by the writings of Charles Darwin in the middle of the previous century, but not applied generally in biology until the twentieth century. Nowadays these ideas permeate many if not all biological disciplines, and the evolutionary framework has been a very productive stimulus to much biological research, especially in behavioral ecology.

This evolutionary spotlight has also been turned on our own species, with analysis of fossils indicating that modern humans are only a few hundred thousand years old, and that our species *Homo sapiens* (wise man) appeared a mere 400,000 years ago.

Genetic manipulation of living organisms has spawned the new disciplines of biotechnology and genetic engineering, both of which are having powerful and far-reaching effects on society, and which will doubtless continue to be central domains of biological advance well into the century ahead. Benefits from these branches of biology include the mass-production of essential drugs, the development of new varieties of crop species, and techniques for detecting, and perhaps also eventually for treating, genetic diseases.

The last decade of the twentieth century has seen the founding and the probable completion of the Human Genome Project (HGP), which aims to map the entire genetic makeup of the human species. This massive research program will hugely increase our understanding of human genetics in general, and of inherited conditions, perhaps helping to promote more effective treatments in the future.

As the twentieth century draws to an end, environmental biologists and ecologists have increasingly stressed the interrelated nature of living organisms and their physical environment, and the general public has become more and more aware of the dangers posed by overpopulation, and by the associated realities of pollution and the destruction of natural ecosystems. Thus biology has a vital role to play in helping modern society develop in closer harmony with nature, rather than simply exploiting the natural world in an unthinking fashion, as tended to be the approach in the previous century.

Arrangement of This Volume

Part I of *Innovations in Biology* comprises six chapters. The first is an introductory essay that presents an overview of biology in the twentieth century and outlines some of the main achievements and initiatives that were responsible for the rapid progress of biological research during this period.

Chapter 2 presents a detailed chronology, which sets out on a timeline key events in biology, such as major discoveries, influential publications, and the dates of foundation of important scientific organizations and institutions.

This chronology is followed in chapter 3 by a selection of sketches of the most famous and influential biologists of the twentieth century, noting the contributions each has made to biology and to the development of their disciplines.

Readers looking for sources of further information will find ample help in chapters 4, 5, and 6, which offer an annotated directory of biology-related organizations, a selected further reading list, and a sampling of Web sites, respectively.

Part II of this volume is a self-contained dictionary that defines more than twelve hundred key biological terms and concepts, some of them illustrated with diagrams for added clarity. This dictionary will enable the reader to find, quickly and conveniently, the meaning of most of the commonly encountered technical terms in biology. This section can also be a useful adjunct to readers' gaining a thorough understanding of discussion in other parts of this handbook. An appendix listing Nobel laureates in the biological sciences through the twentieth century and a general index complete the volume.

Innovations in Biology as a whole is designed to give readers a concise and readable reference to the nature of the biological sciences during the twentieth century, highlighting the most important people, discoveries, and initiatives over that most productive period in the development of the subject. This historical perspective helps to show how biology has developed, and also demonstrates the role that particular scientists and their discoveries have played in shaping modern biology.

Innovations in Biology
Part I

1

Overview

S everal important fields of biology have developed rapidly during the twentieth century, with a number of originally relatively separate disciplines coming together, to the benefit of our overall knowledge.

Some advances have been largely the result of technological improvements, such as the increasing power and resolution of microscopes, culminating in the electron microscope, leading to rapid advances in cell biology. Others have been based more on advances in theoretical guidelines—for example, the strides made in evolutionary studies of behavior and ecology.

Among the more important of these developments have been discoveries in genetics, and in the related fields of developmental and cell biology. For example, we now know that many diseases have a genetic component, and techniques in biotechnology and genetic engineering enable us to detect the responsible genes and to suggest ways in which such diseases may be treated. Genetics has also had a productive impact on ecology and behavioral biology, providing them with a theoretical underpinning that has helped, for example, to explain the evolution of particular strategies in reproduction.

There have been major advances in the study of mammalian, including human, reproduction, with techniques such as in-vitro fertilization becoming almost routine. Prenatal screening for a variety of inherited conditions is also now possible, both for the mother and the unborn child.

Biotechnology and genetic engineering are new biological disciplines, born during the twentieth century, mainly from a fusion of techniques from genetics and biochemistry. These fields are having a significant impact on agriculture and medicine, with new varieties of crop plants being produced by genetic engineering, and techniques perfected for the biological manufacture of drugs.

Two scientists work with an electron microscope in 1941; it was built under the direction of Dr. V. K. Zworykin. (Corbis-Bettmann)

The twentieth century has also seen significant advances in our understanding of evolution as seen through the fossil record, shedding light not least upon the origins of our own species.

Awareness of the global environment has gradually come more and more to the fore throughout the twentieth century, with a realization of the sensitivity of the environment and planet to the destructive effects of pollution.

In all fields of biological research there have been technological advances that have had a considerable impact on progress. Not least has been the rapid rise and development of computers, which have become more and more powerful and portable, and which are now an essential tool in most biological research. The increase in computer speed and analytical power has allowed detailed and very rapid analysis of scientific data, and also the intricate modeling of biological systems.

The Unraveling of Genetics

The first half of the twentieth century was notable for several great steps forward in our understanding of the mechanisms underlying genetics.

The spark for this was actually based on the rediscovery and reinterpretation of much earlier findings, those of the Czech monk Gregor Mendel. Mendel's work on the inheritance patterns of peas was published in 1866, but had little influence on biological thinking until 1900. Then it was scrutinized by Hugo de Vries in Holland, Karl Erich Correns in Germany, and Gustav Tschermak von Seysenegg in Austria. Their work signaled the start of serious research into heredity. Key stages here were the observation and study of genetic mutations, and the discovery of the basic unit of inheritance, the gene. This discovery pointed to a physical mechanism that helped to explain the observed patterns of inheritance of particular traits, such as flower color in plants and eye color in people, to name but two examples.

In 1906, English biologist William Bateson introduced the word "genetics," and in 1909, Wilhelm L. Johanssen, from Denmark, coined the term "gene." Bateson demonstrated that certain inherited traits tend to be transmitted as a group, thus establishing the concept of genetic linkage. In 1909 Bateson used the term "allele" for alternate forms of a gene.

The U.S. biologist Thomas H. Morgan published his influential book *The Theory of the Gene* in 1926. Morgan had many important insights into genetics, and remarked, among other things, upon the phenomenon of sex-linked inheritance and on the significance of the crossing over of chromosomes. Morgan considered the genes, the units of inheritance, to be sited at particular places on the chromosomes, and by determining some of these positions, he began the process of mapping them.

Mention should be made here of the work of the Soviet biologist Trofim Lysenko, who championed the erroneous view that physical traits acquired during life, such as increased muscle mass, could be inherited, basing his ideas partly on the thoughts of the French naturalist Jean Lamarck (1744–1829). Lysenko controlled Soviet biological research from 1928 until 1965, and was strongly supported by the dictator Stalin, who

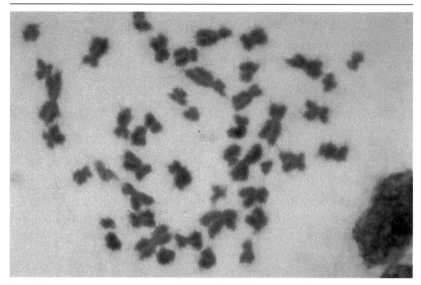

Female human chromosomes magnified four hundred times undergoing mitosis.
(Corbis/Lester V. Bergman)

used Lysenko's theories to underpin and support his own philosophy of
social and political control. In this way an incorrect view of evolution
become enmeshed with an imposed political system, and this view of
evolution affected many decades of biology in the former Soviet Union.

In the mid-twentieth century, the chemical link between the gene and
the expression of that gene in the organism was beginning to be unraveled. This link proved to be the production by the gene of a specific protein, called an enzyme, which has a particular effect in the cell in which it
is produced. In 1941, U.S. scientists George W. Beadle and Edward L.
Tatum established the gene-enzyme hypothesis, showing that one gene
was responsible for the production of one specific enzyme (known famously as "one gene, one enzyme"). Work of this kind was part of the
rapidly developing discipline of biochemistry, which was to become more
and more dominant in twentieth-century biology.

A further piece of the genetic jigsaw puzzle was to fall into place with
the discovery of the chemical nature of the genetic material itself: the
elucidation of the structure of deoxyribonucleic acid (DNA), the compound containing the fundamental genetic information. In 1953, James
Watson and Francis Crick proposed a structural model for DNA, thus
providing a physical mechanism for the storage and transmission of genetic information from one generation to the next. The precise way in
which this genetic information is coded began to be clarified when, in

1961, Marshall W. Nirenberg of the United States cracked the first "letter" of the genetic code.

Research into genetics has continued apace throughout the twentieth century, and much of this research has relied heavily upon a small insect, the fruit fly, *Drosophila melanogaster.* The fruit fly has several advantages that suit it admirably for this role: it is easy to culture in the laboratory, it has a rapid reproductive rate and conveniently large chromosomes, and many of its mutations are easily visible. Such mutations include changes in the shape of legs or wing structures. Mutations, which represent sudden changes in the transfer of inherited characters, began to be seen as playing a central role in the mechanism of evolution.

The New Science of Genetic Engineering

Perhaps the most notable of all areas in which biology has progressed in the twentieth century has been in the field of genetic engineering. This involves altering the genetic makeup of an organism so as to change the type or quantity of the proteins that it produces. Selective breeding of animals and plants has effectively been doing this passively for thousands of years, but genetic engineering now allows scientists to modify an organism's genetic makeup more directly. Genetic engineering relies on certain specific techniques that were discovered this century.

The fruit fly, *Drosophila melanogaster,* is a favorite subject in genetics research. (Corbis/ Robert Pickett)

A key discovery in genetic engineering was made in the by U.S. geneticist Joshua Lederberg in 1952. This was that bacteria come together and exchange genetic material by a process known as conjugation. At about the same time it was realized that many bacteria were evolving resistance to drugs and antibiotics. If resistant bacteria were introduced to a culture of unresistant forms, this resistance could be established rapidly throughout the colony as the resistant genes were spread by the process of conjugation.

Another key stage was the study of viruses that infect bacteria—bacteriophages, or "phages" for short. In 1946, Max Delbrück and Alfred Hershey, working in the United States, showed that genes from phages could combine. Another discovery (by Werner Arber) was that bacteria resist phage attack by using their enzymes to split the phage's DNA. By 1968, Arber had located the enzymes that split the DNA at particular points and found that if the split strands of different genes are put together in the absence of the enzyme they will recombine. The product of this process become known as recombinant DNA, and the process is now a central technique in genetic engineering.

The first researchers to undertake true genetic engineering in this manner were U.S. biochemists Stanley H. Cohen and Herbert W. Brown in 1973. They removed a chunk from the DNA of a bacterium and then inserted a gene from a different bacterium. The engineered bacterium then reproduced, using its altered genetic makeup. Soon scientists had inserted into bacteria not just genes from other bacteria, but genes from other unrelated organisms, such as frogs and fruit flies.

A logical progression from such research resulted in the production of genetically engineered useful products. In 1977, the U.S. scientist Herbert Boyer, of the firm Genentech, fused a segment of human DNA (deoxyribonucleic acid) into the bacterium *Escherichia coli,* allowing it to produce a human protein (in this case somatostatin). This was the first commercially produced genetically engineered product.

Now this technology has moved on, to the point where genetically engineered bacteria and yeasts are used almost routinely to produce useful products such as human growth hormone and insulin. These essential drugs were among the first genetically engineered products to go on sale.

In recent years it has become possible to use genetic engineering techniques to alter the genetic makeup of plants. This offers the possibility of creating new varieties of species, notably crops, with specific new characteristics. For example, resistance to low temperatures can be introduced to crops from warm climates, so they may be grown successfully elsewhere, as can resistance to insect pests and virus diseases. Work is under way with cassava (a staple tropical crop) to introduce disease resistance, and to increase its protein content. Genetically modified crops are now produced in many parts of the world, and are beginning to appear in shops alongside their more "natural" relatives.

The final years of the twentieth century have seen further developments in genetic engineering, many of which have important implications for medicine.

It is now possible to produce what are known as transgenic animals, by injecting genes from one animal into the fertilized egg of another. This technique was first perfected in 1981 by scientists at Ohio University in Athens, Ohio, who injected genes from one animal into the fertilized egg of a mouse. The resulting transgenic mouse then has the foreign gene in many of its cells, and the gene is passed on to its offspring, creating permanently altered (transgenic) animals. In 1982, for example, a gene controlling growth was transferred from a rat to a mouse, producing a transgenic mouse which grew to double its normal size.

Such transgenic techniques promise to benefit medicine, and in particular transplant surgery. One of the major problems in surgery has been in transplanting organs—the main difficulties being the supply of

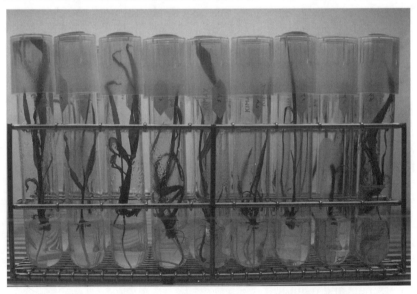

Genetically engineered corn plants grow in controlled conditions for desired genetic traits at the Sungene Technologies Lab in Palo Alto, California. (Corbis/Lowell Georgia)

suitable organs, and the process of rejection by the host. Transgenic pigs have now been bred that incorporate some human genes, which reduces the risk of tissue rejection when the pigs' organs are transplanted into human patients, in a process called xenotransplant.

Parallel research has established that it may soon be possible to grow kidneys in humans that are derived from embryo kidney cells. If successful, these techniques could simultaneously address the twin problems of supply and rejection of human organs. Nevertheless, such research has raised difficult ethical problems, which have yet to be fully resolved.

Many of the rapid advances in animal genetic engineering are of direct medical benefit to humans. Examples of products that can now be manufactured from engineered animals are: human blood-clotting factors, from sheep and goats; human growth hormone and beta-interferon from cattle; and humanized organs for transplant from pigs. Beta-interferon is used to treat multiple sclerosis.

Advances in Human Genetics

In the last few decades of the twentieth century, biological research into human genetics has gathered pace at an impressive rate.

In 1983, Kay Davies and Robert Williamson located the first marker for a human genetic disease, in this case Duchenne muscular dystrophy. This involved studying the DNA from affected and unaffected members of a family, and identifying a piece of DNA that is inherited with the disease. By 1987, the actual gene for this condition was found, and copies were made of it—a process known as gene cloning. By the early 1990s it was found that this gene failed to produce a particular protein that is needed by the skeletal muscles, thus causing the disease. Several other genetic diseases can now be detected by analysis of DNA, in some cases prenatally.

In the mid-1980s, the British geneticist Alec Jeffreys made an important discovery, leading to the development of so-called "genetic fingerprinting." He found that each person carries a unique sequence of short base sequences in their DNA, and that these sequences can be revealed by laboratory analysis, even with only tiny starting samples of DNA. This technique has found a number of uses, such as to resolve cases of disputed parenthood, and to identify criminals from traces of their body fluids, hair, and tissue.

One of the most important developments of genetics is the Human Genome Project (HGP), which is being done in twenty centers around the world, which are in turn coordinated by the Human Genome Organization, or HUGO. This ambitious project, which began around 1990, aims to map and sequence the whole of the human genome—the entire range of genes of the human species—and it is estimated that this will take perhaps fifteen years to achieve. The first stage in this process is the mapping, which is nearing completion. With the help of modern computer power it is hoped that eventually the whole of the biochemical basis of human life will ultimately be unraveled.

One benefit deriving in part from this effort is the detection of inherited genetic disorders and diseases, and of people who may be carriers of such conditions, even in the unborn embryo during pregnancy. The HGP will also improve our understanding of genetic diseases, such as cystic fibrosis, and perhaps pave the way toward more effective treatment. Work on the HGP has continued to gather speed, and it now looks as though the millennium may see the achievement of the project's goal, the complete sequencing of the three billion base-pairs that make up the entire human genome.

The Birth of Biochemistry and Cell Biology

During the twentieth century, biology gained enormously by absorbing the findings of chemists, and in particular from the study of organic compounds, which provide the building materials of all living organisms. This led to the rise of a new discipline, biochemistry, concerned with the chemistry of living organisms.

Improvements in the quality and power of lenses led also to advances in the field of cell biology. A major step forward here was provided by the invention of the electron microscope, which revealed structures previously invisible to the human eye, enabling even intra-cellular structures to be examined clearly.

Throughout the twentieth century biologists became increasingly aware of the role of specific chemicals in controlling bodily processes. The first such chemical to be discovered, and then synthesized, was the heart stimulant adrenaline (epinephrine). Adrenaline was discovered by the Japanese-born U.S. biochemist Jokichi Takamine in 1901. This was soon followed, in 1902, by the discovery of peristalsis and of the hormone secretin by British physiologists Ernest Henry Starling and William M. Bayliss.

By the turn of the century, people began to realize that soil, although it is full of bacteria and fungi, does not generally cause diseases. Something in the soil seems to stop disease-producing bacteria from thriving there, and scientists began to search for this magic factor. In 1922, Scottish bacteriologist Alexander Fleming isolated a substance, called lysozyme, from tears, which he demonstrated could kill bacteria. Then in 1928, an accidental event in his laboratory led to a major biological discovery that was to have a huge impact. A mold (a type of fungus) had grown on a petrie dish in which there was a culture of a bacterium, *Staphylococcus*. Around each speck of mold was a ring of dead bacteria. Fleming isolated a substance from the mold and found that it killed certain bacteria, but not others, nor did it damage white blood cells. Thus was penicillin discovered.

Eventually, methods were perfected for extracting large quantities of penicillin from cultures of mold, and this drug was then used to save thousands of lives during World War II. Since 1943, other natural antibiotics, such as streptomycin and aureomycin, have been discovered.

More recently, many more antibiotics have been synthesized by biochemists. However, although they are an effective way of combating infection, the liberal use of antibiotics has promoted the evolution by the target bacteria of resistant strains, and this is posing an increasing problem for disease control in the latter years of the twentieth century.

Developmental Biology

In developmental biology, great strides forward were also made, the main one being the realization that all the cells of an organism undergo a process of differentiation, with only a small fraction of the genes in a single cell being active at a given time.

Around 1900, the distinction was made between tissues that seem to develop according to an intrinsic genetic program (determinate development) and those that are affected by adjacent tissues or structures (regulative development). This led to the new concept of "induction," where one tissue affects the development of another. Induction was first shown clearly by German embryologist Hans Spemann (1901), using the eye in a frog embryo. These early findings in developmental biology have continued through the century and gradually the position-dependent development of cells is being teased apart. In this new field of topobiology, the cell membranes are being investigated ever more closely, to decipher the precise mechanisms involved.

A marriage between genetics and developmental biology came about with the demonstration, by Canadian-born U.S. bacteriologist Oswald Avery in 1944, that DNA was the carrier of the genetic information, controlling the production of the proteins forming the organism. Development consists of the marshaling of these proteins, and their combination to form the different organ systems in the body, in an ordered sequence of gene expression.

Throughout the twentieth century there has been a steady increase in our knowledge of the processes controlling reproduction, and in particular human reproduction. Oral contraceptives, first produced in the mid-1950s, have been refined and are now routinely used by millions of women as an effective means of family planning. Globally, this control has helped to begin to stem the problem of overpopulation, arguably the greatest problem to face humanity, with the effects it brings with it of famine and poverty, and its impact on our environment in general.

A major event in reproductive biology, this time in the late 1960s, was the successful fertilization of human egg cells outside the body—the so-called test-tube baby technique, or in-vitro fertilization. 1978 saw the first baby, Louise Brown, born using this technique. Having been unable to remove a blockage from her mother's Fallopian tube, English obstetrician Patrick Steptoe and Welsh physiologist Robert Edwards removed an egg from her ovary, fertilized it with her husband's sperm, and implanted it in her uterus.

Since the 1970s there have been several further advances in our knowledge of reproduction, including the development of some controversial techniques. It may be possible in the near future to allow fetal development to continue outside the mother's body. In a key experiment, in 1997, the Japanese scientist Yoshinori Kuwabara announced that his team had successfully grown goat embryos in an acrylic tank. The embryos, which were removed from their mother at seventeen weeks into pregnancy, were placed in the tank, which was filled with liquid simulating amniotic fluid. A machine that pumped oxygen and nutrients into the embryo's blood replaced the role of the placenta. At twenty weeks' gestation, the goat was "born." At present the procedure can only be done late in fetal development. Were it to be extended to people, it could be used to rescue babies who become at risk early in a pregnancy, or who are born very prematurely.

The final decade of the twentieth century has also seen the successful cloning of animals, first undertaken with a rabbit in 1975. Eventually, in 1996, two monkeys were successfully cloned from embryo cells, indicating that cloning techniques would probably be possible in humans as well.

In 1997, British geneticists successfully cloned an adult sheep. In this process, a cell was taken from the udder of the mother sheep and its DNA combined with an unfertilized egg that had had its DNA removed. The fused cells were grown in the laboratory and then implanted into the uterus of a surrogate mother sheep. The resulting lamb, Dolly, became famous as the first animal cloned using cells other than reproductive cells.

Theoretically at least, a similar process could clone members of our own species, and this realization has resulted in international calls to prevent the cloning of humans. Despite the potential benefits of cloning technology, there are major concerns about the ethics of extending its use to humans. The artificial production of cloned humans may be possible, but raises obvious moral problems. In June 1997, the U.S. National Bioethics Advisory Commission proposed a five-year ban on human cloning, and in 1998, nineteen European countries signed an international ban on human cloning.

Ecology and Animal Behavior

Ecologists study animals and plants in their habitats, and much of this involves seeing how populations change with changing conditions. The term *ecology* was coined in the 1860s, by German zoologist Ernst Haeckel, for the study of the interrelations of animals and plants. But it is only in the twentieth century that ecology developed into an active field of biological research.

In the eighteenth and nineteenth centuries, ecology (or natural history as it was then known) was mainly a descriptive cataloging of God's ordered natural world, and no need was felt to search for underlying mechanisms. In the twentieth century, however, the major innovation in ecology was that its investigations become explanatory—biologists began to search for reasons to explain their observations, by experiment and by the testing of theories. As elsewhere in biology, the work of English naturalist Charles Darwin, whose *The Origin of Species* prompted a storm of controversy when it was published in 1859, had a huge influence on ecology and gave it a secure theoretical support. Another big influence here was the publication in 1927 of British ecologist Charles Elton's *Animal Ecology,* which set out a population biology approach to ecology.

Before the rise of the computer, such studies were difficult and time-consuming, but the increase in power and sophistication of these machines has allowed ecologists to analyze much more data and with far greater accuracy, and to model the behavior of natural populations and compare these models with reality.

Many ecological field studies investigate how different species compete with each other for resources and have shown differences in reproductive strategy, related to environment and habitat. Ecologists have also been responsible for unraveling some of the many complex feeding relations, or food chains, highlighting the interdependence of organisms in nature.

Effective ecological work is impossible without a sound knowledge of the identities of the species involved. In the nineteenth century, taxonomy (the classification of life forms) and natural history tended to go hand in hand, but in recent decades, species description and naming have often come to be regarded as unglamorous pursuits. Nevertheless, it is essential to increase our knowledge in this area and to attempt to expand our inventory of species, especially now so many of them are under threat. It is a sobering thought that despite the efforts of taxonomists, there is still no reliable estimate of the number of species that exist today. To take the case of animals, about 1.5 million species have been described, of a minimum of some 10 million species estimated to exist, representing just 15 percent.

Advances in our understanding of genetics and of evolution also had a significant impact on ecology, resulting in the disciplines of evolutionary and behavioral ecology. In particular, the application of the study of genetics and inheritance proved powerful in explaining and predicting the behavior of many species of animals.

The incorporation of evolutionary thinking into studies of animal behavior gave rise to the science of sociobiology, heralded by the publication of *Sociobiology: The New Synthesis* by U.S. zoologist Edward O. Wilson in 1975.

Wilson and others showed that many different patterns of social behavior in animals can be explained in terms of the evolution of behavior "designed" to promote the survival of their genes. This analytical method proved particularly powerful in explaining the many examples of altruistic behavior, in which individuals behave, or appear to behave, in an unselfish manner. Close study has revealed that in many such cases the danger the altruistic individual appears to risk may be offset by other advantages. In other cases, individual behavior may be truly altruistic, but the beneficiaries are nearly always closely related genetically.

During the twentieth century there have been many important field studies that have carefully observed the behavior of animals in the wild. In particular the observations of chimpanzees and their close relative the bonobo (or pygmy chimpanzee) have shed a great deal of light on the complex patterns of social behavior of these primates. Among the most important of such studies have been the detailed work on wild chimpanzees by the English primatologist and conservationist Jane Goodall since

1960. Thanks to this research we now know that chimpanzees have highly complex and elaborate social lives, with many parallels to human behavior, and that they are omnivorous, not strict vegetarians as once thought. Similar studies were made by U.S. zoologist Dian Fossey, this time on gorillas in Rwanda, revealing these powerful apes to have peaceful, family-based lives, browsing on forest vegetation. Goodall and Fossey also pointed out the growing need for effective conservation policies to protect the great apes—Dian Fossey died at the hands of illegal poachers, whose traps she had interfered with, becoming a martyr to the cause of conservation.

The Environment

The twentieth century saw the birth of awareness about the need to care for our environment and of the destructive effects of pollution. Influential here was the publication in 1962 of the book *Silent Spring,* by U.S. ecologist Rachel Carson, which warned of the dangers chemical pesticides pose to humans and the environment, and which perhaps more than any other event launched the environmental movement.

This awareness has continued to gather pace through the twentieth century as we have realized more and more that the earth is a closed system with its own often delicate balance of physical and biological factors supporting the living organisms that thrive here and in turn influence its makeup.

This whole-earth view gained its most famous expression with the formulation of the Gaia hypothesis, by British scientist James Lovelock, in the mid-1960s. This view regards the planet earth as a complex, self-regulating system, easily disturbed, but able to adjust, almost like a giant single organism.

The Gaia concept has undoubtedly helped focus peoples' thoughts on the importance of a global ecological view of our planet and has highlighted the damage caused by human activity on the natural ecological pathways, damage that potentially threatens the future of the earth itself. And while the Gaia hypothesis has helped to focus biological thought, it has also attracted a less critical, almost mystical, following, perhaps because of its name, after an ancient Greek earth goddess.

In recent decades, environmental ideas have become established, and have now started to feed through into a new environmental ethic that recognizes our responsibility for the rest of the natural world. In 1972, the United Nations Conference on the Human Environment was held in Stockholm, Sweden. This was the first international conference on the environment, and its stated aim was to improve the world's environment through monitoring, resource management, and education.

Put at its starkest, the problem faced by human society today, and recognized clearly for the first time during the twentieth century, is a conflict between individual freedom and the health of the planet. Unsustainable exploitation of the natural world and overpopulation are two of the biggest problems we face, and perhaps the best hope for the future lies in a view of life that recognizes our responsibilities and states that we as individuals should strive not to do anything to our environment that would make life more difficult for future generations. In essence, we all share a responsibility for the future of our species, and also for the diversity of life on the whole planet. A true and deep understanding of biology, of evolution, and of our place in nature leads naturally to this worldview which, if widely implemented, gives us the best hope of a healthy future.

Particularly dangerous among the threats to the environment are global warming, the destruction of the rain forests, and human-induced desert expansion.

During the final decades of the twentieth century, biologists have become increasingly concerned about the effects of pollution, particularly on the atmosphere. It is now well established that chemical pollution in the at-

An area of the Malaysian rain forest being burned for agricultural and construction purposes, 1990. (Corbis/Sally A. Morgan; Ecoscene)

mosphere has thinned the naturally occurring layer of high-level ozone. The latter protects the earth from much harmful radiation. At the same time, accumulated pollutants, notably carbon dioxide, have increased the greenhouse effect by trapping heat in the atmosphere, threatening a warming of global average temperatures. Scientists are now broadly in agreement that global warming (that is, human-induced recent global warming) is a reality. The dangers are the unpredictable effects of alterations to global climate patterns, but also the raising of sea levels caused largely by a melting of polar ice. Average world temperatures are expected to have risen by 1.5°C/34.7°F by the year 2025, with a 20cm/7.8 in rise in sea level.

The Elucidation of Evolution and of Human Origins

Even though earlier biologists, notably Lamarck in 1809, had postulated the descent of humans from primates, it was not until the last decades of the nineteenth century that evolution began to enter into the mainstream of biological thought and research. Since then it has become very much a central skeleton supporting many varied branches of biological research and spawning disciplines such as human biology, rooted in physical and cultural anthropology, as well as anatomy, physiology, and genetics.

Thus, in the twentieth century, our own species, although obviously unique and special, is no longer regarded as separate from the rest of life and evolutionary processes, and biologists have increasingly applied evolutionary principles to many different areas of their research, perhaps most notably in ecology and animal behavior.

In 1859, with the publication of *On the Origin of Species by Means of Natural Selection or the Preservation of Favored Races in the Struggle for Life* (*The Origin of Species,* for short), Darwin revolutionized our understanding of evolution. But it was not until the early 1930s that the findings of the geneticists and evolutionary biologists began to be reconciled. In 1930, the English geneticist Ronald Fisher published *The Genetical Theory of Natural Selection,* in which he synthesized Mendelian genetics and Darwinian evolution. This synthesis was also due in no small part to the Ukraine-born U.S. scientist Theodosius Dobzhansky, who was both naturalist and geneticist. In 1937 he distilled his thinking in the seminal work *Genetics and the Origin of Species.*

This century has seen major discoveries of human fossils, which have helped to shed light on the evolution of our own species. Around the turn of the century, fossils of a species called *Homo erectus* (Java Man) were discovered. This humanlike creature (hominid) lived between 0.5 and 1.5 million years ago. In the 1920s and 1930s, an even older form was discovered, and named in a separate genus *Australopithecus.* This hominid lived between about 1 million and 2 million years ago.

Since the 1950s, even older forms of *Australopithecus* have come to light, in Africa. More recently, in 1994, *A. ramidus* has been dated at 4.4 million years old. This form was probably mainly vegetarian and lived in a woodland habitat. Like humans, it walked upright on its hind legs.

In the 1960s and 1970s, a form regarded as intermediate between the modern human *(Homo sapiens)* and *Australopithecus* was found in Africa. In 1964, this creature was named Handy Man *(Homo habilis)* by Louis Leakey, Philip Tobias, and John Napier, after its apparent use of stone tools. The fossils came from the now famous Olduvai Gorge in Tanzania, and have been dated at between 1.6 million and 2 million years old.

Various further finds of *Homo erectus* (named for its humanlike upright stance) have been made, in Europe and Africa, mainly in northern Kenya, but also in Algeria, Morocco, Ethiopia, and South Africa. The most likely story emerging is that this form originated in Africa, then spread to Europe and Asia. *Homo erectus* probably used tools made of wood, such as hand axes, and were hunters and gatherers with a mixed diet.

The techniques of genetics and biochemistry were also applied to physical anthropology, and analyses of protein and DNA from fossils and living humans and apes were undertaken beginning in the early 1990s. These studies revealed that modern humans are more closely related to chimpanzees and bonobos (pygmy chimpanzees) than either are to gorillas, gibbons, and orangutans. They also helped shed light on the evolutionary pathways on the fossil hominid line.

Analysis of DNA and fossils suggests that modern humans are only a few hundred thousand years old, and that the evolutionary split between humans and apes occurred between 5 million and 8 million years ago, probably in Africa. Our own species, *Homo sapiens* (Wise Man) emerged as recently as around 400,000 years ago. All living people are closely related and probably share a common ancestor in Africa.

A form of human, *Homo neanderthalensis,* known as Neanderthal Man (from the Neander valley in Germany where it was first discovered) evolved in Europe around 200,000 years ago. A genetic analysis carried out on DNA extracted from fossil Neanderthal bones indicated in 1997 that Neanderthals shared a common ancestor with modern humans no later than 600,000 years ago and proved conclusively that they were not our direct ancestors but represent a separate evolutionary line.

The influx of evolutionary thinking into human biology has not been without its problems, however, and most notoriously these have been applied to aspects of racial differences, and in the pseudo-science of eugenics. Since humans have so clearly evolved, can the "quality" of the species be improved? This is a question that has been asked by many people, and distorted sometimes for sociopolitical ends, perhaps most famously by Adolf Hitler. It was Darwin's cousin Sir Francis Galton who first coined the term *eugenics*, suggesting that selection be used to improve humankind to the benefit of future generations of people. Sadly, most attempts to do so have been disastrous and misguided. There are a number of reasons why eugenics is undesirable, not least the undemocratic nature of the processes that would be required—who is to decide which individuals deserve to contribute their genes to the future? We now realize that such aims are also not beneficial, and that to function effectively human society needs a rich mixture of individuals of different genetic makeup. Thankfully perhaps, eugenic improvement is probably impossible, and is now largely discredited.

Modern human society operates largely outside the constraints of natural selection, and the human gene pool is highly variable. Nevertheless, our increasing knowledge of genetically based disorders does allow voluntary decisions to be taken, for example by parents who may be carrying certain genes that increase markedly their likelihood of producing a disabled child. This could be seen as a kind of voluntary, limited form of eugenics.

Extinction and the Discovery of New Species

More than 99.9 percent of all evolutionary lines that once existed on earth have become extinct, so extinction has loomed large in the history of life on earth and continues to feature in the natural world today. On the other hand, our knowledge of the natural world remains so sparse that species are still being discovered and described.

Discoveries by paleontologists and geologists have helped to explain the apparently catastrophic extinctions that seem to have occurred from time to time in the distant past. The most famous of these was the relatively sudden demise of the dinosaurs toward the end of the Cretaceous period.

Analysis of a clay layer in deposits dated at the boundary of the Cretaceous and Tertiary periods by U.S. father-and-son physicists Walter and Luis Alvarez in 1980 revealed unusually rich traces of the heavy metal iridium. It was then found that this enriched iridium layer was present in deposits of similar age worldwide. The Alvarezes proposed the theory that this was caused by the impact on the earth of a large asteroid, which

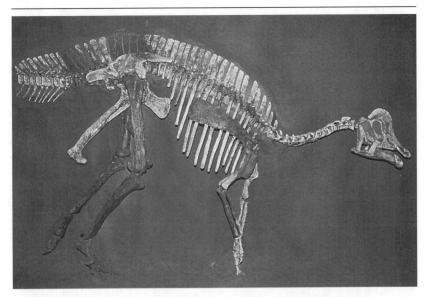

A Hypacrosaurus fossil from the Cretaceous era. (Corbis/Kevin Schaefer)

might have had other ecological effects and possibly explain the sudden extinction of many creatures, including the dinosaurs.

David Raup and John Sepkoski, Jr., looked at other mass extinctions of the past, and found that these were periodic, happening around every twenty-six million years. They developed the theory that something periodically disturbed the cloud of comets (known as the Oort cloud) at the edge of our solar system, causing some of the comets to fall toward the sun, occasionally hitting earth or other planets.

As we reach the end of the twentieth century, new species of animals and plants are still being discovered and described. Expeditions to tropical rain forests routinely find undescribed insects, notably beetles, but there have also been recent discoveries of new vertebrates, even some quite large mammals. For example, a new species of whale was described from Chile in 1996, and a new species of muntjac deer from Vietnam in 1997.

How Life on Earth Began

The problem of how life began on earth has intrigued biologists for centuries, but in the twentieth century some of the mystery of this process has been removed.

In the early 1900s, Swedish scientist Svante Arrhenius suggested that life may have been seeded from elsewhere in the universe (panspermia), and this idea was revived most notably by British astronomer Fred Hoyle in the 1970s. In the 1950s it was becoming clear that the early atmosphere of the earth probably consisted mainly of methane, ammonia, hydrogen, and water, but with little free oxygen. By passing electric current through this mixture in the laboratory, U.S. chemist S. L. Miller produced at least twenty-five different amino acids, the precursors of the proteins that are so central to living things. However, the discovery of DNA as the central and probably in some sense ancestral molecule of life made it hard to see how life could have evolved from a soup of these early chemicals. How could this mixture organize itself into the complexity required for replication? Most biologists now agree that RNA (a simpler form of DNA) probably formed at some stage and led to the evolution of simple replicating organisms, although no one has yet explained how this might have happened.

So, although advances have been made this century, speculation remains on this fundamental topic in biology. What we do know is that the earliest forms of life were simple, single-celled bacteria, dating from 3.5 billion years ago. From these organisms, the whole of the rest of life on Earth gradually evolved. Although the mechanism and possible routes for evolution are generally agreed and understood, we still do not know precisely how the whole process began.

—*Martin Walters*

2

Chronology

1900

The Galton-Henry fingerprint classification system is published. It is adopted by the English police force at Scotland Yard in London in 1901; other law enforcement agencies change over soon after.

Austrian immunologist Karl Landsteiner discovers the ABO blood group system.

Dutch geneticist Hugo Marie de Vries, German botanist Carl Erich Correns, and Austrian botanist Erich Tschermak von Seysenegg simultaneously and independently rediscover the Austrian monk Gregor Mendel's 1865 work on heredity.

The Kral Collection of microorganisms is established in Prague, Czechoslovakia; it is the first collection of pure cultures of microorganisms used for research purposes.

Dutch biologist Samuel Schouten describes a method of isolating a single bacterium in the field of a high-powered microscope.

1901

English biochemist Frederick Gowland Hopkins isolates the amino acid tryptophan.

Japanese-born U.S. biochemist Jokichi Takamine first synthesizes the heart stimulant epinephrine (adrenaline) from the suprarenal gland. It is the first pure hormone to be synthesized from natural sources.

1902

British physiologists William Bayliss and Ernest Starling discover that a substance, which they call secretin, is released into the bloodstream by cells in the duodenum. It stimulates the secretion of digestive juices by the pancreas and is the first hormone to be discovered.

French physiologist Charles Richet discovers cases of acute sensitiveness to antidiphtheria serum, which he calls "anaphylaxis." His work leads to a greater understanding of problems of asthma, hay fever, and other allergic reactions.

German chemists Emil Fischer and Franz Hofmeister discover that proteins are polypeptides consisting of amino acids.

1902–1904

U.S. geneticist Walter Sutton and the German zoologist Theodor Boveri put forth the chromosomal theory of inheritance when they show that cell division is connected with heredity.

1903

Russian physiologist Ivan Pavlov describes learning by conditioning. He trains dogs to expect food when they hear a bell and eventually they salivate every time the bell rings.

1904

Spanish physiologist Santiago Ramón y Cajal demonstrates that the neuron is the basis of the nervous system.

1905

Danish botanist Wilhelm Johannsen introduces the terms "genotype" and "phenotype" to explain how genetically identical plants differ in external characteristics.

English physiologists William Bayliss and Ernest Starling coin the word "hormone" (from Greek hormon, "impel") to describe chemicals that stimulate an organ from a distance.

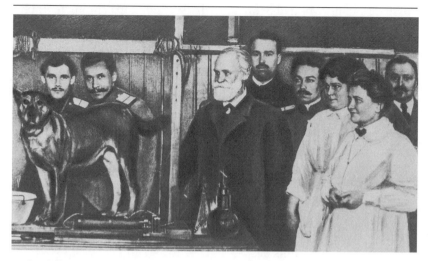

Undated illustration of Ivan Pavlov and his staff demonstrating the conditioned reflex phenomenon. (Corbis-Bettmann)

Scottish physiologist John Scott Haldane discovers that breathing is regulated by the concentration of carbon dioxide in the blood affecting the respiratory center of the brain.

c. 1905

English biochemist Frederick Gowland Hopkins shows that the amino acid tryptophan and other essential amino acids cannot be manufactured from other nutrients but must be supplied in the diet.

1906

British biochemists Arthur Harden and William Young discover catalysis among enzymes.

English biologist William Bateson introduces the term "genetics."

English neurophysiologist Charles Scott Sherrington publishes *The Integrative Action of the Nervous System,* in which he classifies sense organs into three main groups: exteroceptive (sight, smell, hearing, and touch), interoceptive (taste), and proprioceptive (interior receptors that control balance and breathing and so on).

**GIBSONS AND DISTRICT
PUBLIC LIBRARY**

Russian botanist Mikhail Semyonovich Tsvet develops chromatography for separating plant pigments.

1907

English biochemists Frederick Gowland Hopkins and Walter Fletcher show that working muscle accumulates lactic acid, which leads to a greater understanding of the chemistry of muscular contraction.

German chemist Emil Fischer publishes *Researches on the Chemistry of Proteins*, in which he describes the synthesis of amino acid chains in proteins.

Scottish physiologist John Scott Haldane develops a stage-decompression method that permits deep-sea divers to ascend safely.

U.S. zoologist Granville Ross Harrison develops the first successful animal tissue cultures; they prove vital in cancer research.

1908

English mathematician Godfrey Hardy and German physician Wilhelm Weinberg establish the mathematical basis for population genetics, the Hardy-Weinberg principle.

1909

Danish botanist Wilhelm Ludvig Johannsen introduces the term *gene*.

English biologist William Bateson publishes *Mendel's Principles of Genetics*, which introduces Mendelian genetics to the English-speaking world.

German botanist Carl Correns shows that certain hereditary characteristics of plants are determined by factors in the cytoplasm of the female sex cell. It is the first example of non-Mendelian heredity.

Russian-born U.S. chemist Phoebus Levene discovers D-ribose, the five-carbon sugar that forms the basis of RNA.

1910

U.S. geneticist Thomas Hunt Morgan discovers that certain inherited characteristics of the fruit fly *Drosophila melanogaster* are sex-linked. He later argues that because all sex-related characteristics are inherited together, they are linearly arranged on the X chromosome.

1912

English biochemist Frederick Gowland Hopkins publishes the results of his experiments that prove that "accessory substances" (vitamins) are essential for health and growth and that their absence may lead to diseases such as scurvy or beriberi.

English physiologist Ernest Starling publishes *Principles of Human Physiology*, which is still an international standard text on physiology.

Polish-born U.S. biochemist Casimir Funk discovers that pigeons fed on rice polishings can be cured of beriberi and suggests that the absence of a vital nitrogen-containing substance known as an amone causes such diseases. He calls these substances "vitamines."

U.S. entomologist Leland Ossian Howard publishes *The House Fly, Disease Carrier*, in which he identifies the common housefly as a major carrier of disease.

1913

German chemist Richard Willstätter determines the composition of chlorophyll.

U.S. biochemist Elmer Verner McCollum isolates vitamin A.

U.S. physiologist John Abel invents the first artificial kidney.

1914

German biochemist Fritz Albert Lepmann explains the role of adenosine triphosphate (ATP) as the carrier of chemical energy from the oxidation of food to the energy consumption processes in the cells.

U.S. biochemist Edward Kendall isolates the hormone thyroxine from the thyroid gland. It regulates metabolism by stimulating all cells to consume oxygen.

1915

U.S. geneticists Thomas Hunt Morgan, Alfred Sturtevant, Calvin Bridges, and Hermann Muller publish *The Mechanism of Mendelian Heredity*, which outlines their work on the fruit fly *Drosophila* demonstrating that genes can be mapped on chromosomes.

1917

U.S. researcher D. F. Jones discovers the "double cross" technique of hybridizing corn: four inbred lines, instead of two, are crossbred, resulting in increased yields making commercial production more practical.

1918

Scottish geneticist Ronald Fisher shows that both genes and environmental factors affect an individual's behavior.

1919

Austrian zoologist Karl von Frisch discovers that bees communicate the location of nectar through wagging body movements and rhythmic dances.

1920

Russian botanist Nikolay Ivanovich Vavilov states that a plant's place of origin is the region where its greatest diversity is found. He identifies twelve world centers of plant origin.

1921

Canadian microbiologist Félix-Hubert D'Hérelle publishes *Le Bactériophage, son rôle dans l'immunité [The Bacteriophage, Its Role in Immunity]*, in which he describes the discovery of bacteriophages, viruses that infect bacteria.

Scottish bacteriologist Alexander Fleming discovers the antibacterial enzyme lysozyme, which is found in tears and saliva.

Canadian physiologists Frederick Banting, Charles Best, and John MacLeod isolate insulin. A diabetic in Toronto receives the first insulin injection.

U.S. botanist Edward Murray East develops a high-yield hybrid corn.

1922

English biochemist Frederick Gowland Hopkins isolates glutathione and demonstrates its vital role in the cell's utilization of oxygen.

French surgeon Alexis Carrel discovers white blood cells (leukocytes).

U.S. chemist Herbert McLean Evans discovers vitamin E.

1923

French bacteriologists Albert Calmette and Camille Guérin develop the tuberculosis vaccine, known as Bacillus Calmette-Guérin (BCG), and use it to vaccinate newborns at a hospital in Paris, France.

1924

English physiologist Ernest Starling finds that bicarbonates, chlorides, glucose, and water excreted by the kidney are reabsorbed by the glomeruli at the lower end of the kidney tubules.

1925

July 10–July 21. The celebrated "Scopes monkey trial" is held in Dayton, Tennessee. The case pits liberal lawyer Clarence Darrow against politician and fundamentalist William Jennings Bryan in the case of a schoolteacher, John T. Scopes, who is being convicted for teaching the theory of evolution contrary to state law. Scopes is convicted and fined $100, but this is waived on a technical point.

Clarence Darrow leans on a desk at the famous Scopes trial in Dayton, Tennessee. John T. Scopes stands behind him with his arms crossed. (Corbis-Bettmann)

English geneticist Ronald Fisher publishes *Statistical Methods for Research Workers*, in which he demonstrates experimental techniques and statistical methods to be used in biology.

U.S. anatomist Florence Sabin becomes the first woman member of the American Academy of Sciences.

U.S. geneticists Thomas Hunt Morgan, Alfred Sturtevant, and Calvin Blackman Bridges publish the results of their genetic experiments with the fruit fly *Drosophila melanogaster*, showing that genes can be mapped onto chromosomes.

U.S. pathologist George Whipple demonstrates that iron is the most important factor involved in the formation of red blood cells.

1926

U.S. biochemist Elmer McCollum isolates vitamin D and uses it to treat rickets successfully.

U.S. biochemist James Sumner crystallizes the enzyme urease. It is the first enzyme to be crystallized. Sumner's achievement demonstrates that enzymes are proteins.

U.S. geneticist Thomas Hunt Morgan publishes *The Theory of the Gene*, in which he demonstrates that the gene will form the foundation of all future genetic research.

U.S. physiologist John Jacob Abel isolates and crystallizes insulin.

1927

Austrian-born U.S. immunologist Karl Landsteiner discovers the M and N blood groups.

U.S. geneticist Hermann Muller uses X-rays to cause mutations in the fruit fly. His work permits a greater understanding of the mechanisms of variation.

1928

English physiologists Edgar Douglas Adrian and Charles Sherrington publish *The Basis of Sensation*, which discusses how the nerves transmit messages to and from the brain.

English bacteriologist Fred Griffith discovers that the virulence of pneumococci bacteria (responsible for pneumonia) depends on the presence of an envelope of polysaccharides (sugar units) surrounding the bacteria cells.

U.S. biochemist Charles King and Hungarian biochemist Albert Szent-Györgyi independently discover vitamin C.

1929

British neurologist Edgar Douglas Adrian, using an ultrasensitive galvanometer, is able to follow a single impulse in a single nerve fiber, an achievement that aids understanding of the physical basis of sensation.

German biochemist Adolf Butenandt and, simultaneously and independently, the U.S. biochemist Edward Doisy isolate the hormone estrone, which is involved in the growth and development of females.

The Migratory Bird Conservation Act in the United States provides authority and funds for the establishment of refuges for migrating birds.

1930

Swiss biochemist Paul Karrer formulates the structure of betacarotene, the precursor to vitamin A.

English geneticist Ronald Fisher publishes *The Genetical Theory of Natural Selection,* in which he synthesizes Mendelian genetics and Darwinian evolution.

U.S. biochemist Edward Doisy crystallizes the hormone estriol, the first estrogen hormone to be crystallized.

U.S. biochemist John Northrop crystallizes pepsin and trypsin, demonstrating that they are proteins.

1931

German biochemist Adolf Butenandt isolates the male sex hormone androgen.

U.S. biologist Ernest Goodpasture grows viruses in eggs, making possible the production of vaccines for such viral diseases as polio.

1932

In Germany, biochemist Hans Krebs discovers the urea cycle, in which ammonia is turned into urea in mammals.

1933

British geneticist John B. S. Haldane popularizes evolution with the publication of *The Causes of Evolution.*

Canadian biologist Ludwig von Bertalanffy writes *Theoretical Biology* in which he attempts to develop a common methodological approach to all sciences based on the tenets of organismic biology.

1934

German biochemist Adolph Butenandt isolates the female sex hormone progesterone.

Norwegian biochemist Asbjörn Fölling discovers the genetic metabolic defect phenylketonuria, which can cause retardation; his discovery stimulates research in biochemical genetics and the development of screening tests for carriers of deleterious genes.

1935

Austrian zoologist Konrad Lorenz founds the discipline of ethology, the comparative study of animal behavior in its natural setting, by describing the learning behavior of young ducklings; visual and auditory stimuli from the parent object cause them to "imprint" on the parent—the process by which they become instinctively attached to the parent.

U.S. biochemist Edward Calvin Kendall isolates the steroid hormone cortisone from the adrenal cortex.

U.S. biochemist Wendell Meredith Stanley shows that viruses are not submicroscopic organisms but are proteinaceous in nature.

1937

French microbiologist Max Theiler develops a vaccine against yellow fever; it is the first antiviral vaccine.

German-born British biochemist Hans Krebs describes the citric acid cycle in cells, which converts sugars, fats, and proteins into carbon dioxide, water, and energy—the "Krebs cycle."

1938

December. A coelacanth, an ancient fish assumed to be extinct, is discovered in the Indian Ocean.

1939

U.S. microbiologist René J. Dubos is the first to search systematically for, and discover, natural antibiotics. He looks for soil bacteria that kill other bacteria and discovers the antibiotics gramicidin and tyrocidine.

1940

Working in the United States, Austrian-born U.S. immunologist Karl Landsteiner and U.S. physician and immunohematologist Alexander Wiener discover the rhesus (Rh) factor in blood.

The globulin, albumin, and fibrin fractions in blood are separated by U.S. scientist Edwin J. Cohn. Albumin is used to treat shock, globulin to prevent infection, and fibrin to stop hemorrhaging.

U.S. microbiologist Thomas Francis, Jr., isolates the virus responsible for influenza B.

U.S. physiologist Herbert M. Evans uses radioactive iodine to prove that iodine is used by the thyroid gland.

1944

British chemists Archer J. P. Martin and Richard L. M. Synge separate amino acids by using a solvent in a column of silica gel. The technique represents the beginnings of partition chromatography and leads to further advances in chemical, medical, and biological research.

Hungarian-U.S. mathematician John von Neumann publishes *The Theory of Games and Economic Behavior*. Games theory becomes an important tool in the study of animal behavior.

The role of deoxyribonucleic acid (DNA) in genetic inheritance is first demonstrated by U.S. bacteriologist Oswald Avery, U.S. biologist Colin MacLeod, and U.S. biologist Maclyn McCarthy; the discovery opens the door to the elucidation of the genetic code.

1946

U.S. biologists Max Delbrück and Alfred D. Hershey discover recombinant DNA (deoxyribonucleic acid) when they observe that genetic material from different viruses can combine to create new viruses.

U.S. geneticists Joshua Lederberg and Edward Lawrie Tatum pioneer the field of bacterial genetics with their discovery that sexual reproduction occurs in the bacterium *Escherichia coli.*

1947

English physiologists Alan Hodgkin and Andrew Huxley insert microelectrodes into the giant nerve fibers of the squid *Loligo forbesi* to discover the chemical and electrical properties of the transmission of nerve impulses.

The U.S. educator and agricultural chemist Karl Paul Link develops the rat poison warfarin; an anticoagulant, it causes rats to bleed to death.

1948

U.S. biologist Alfred Mirsky discovers ribonucleic acid (RNA) in chromosomes.

Scottish geneticist Charlotte Auerbach's studies begin the science of chemogenetics.

Soviet biologist Trofim D. Lysenko outlaws orthodox genetics in favor of the belief in acquired characteristics in the USSR. Purges of orthodox geneticists (and assertions that, for example, wheat plants can produce rye seeds) obstruct agricultural development.

Swiss physiologist Walter Hess describes using fine electrodes to stimulate or destroy specific regions of the brain in cats and dogs; the technique allows him to discover the role played by various parts of the brain.

1949

U.S. researchers synthesize adrenocorticotropic hormone (ACTH), which the pituitary gland secretes to stimulate the adrenal glands.

1951

U.S. biochemist Robert Woodward synthesizes cortisone.

1952

Austrian zoologist Konrad Lorenz publishes *King Solomon's Ring*, in which he argues that natural selection works on behavioral as well as physical characteristics.

English biophysicist Rosalind Franklin uses X-ray diffraction to study the structure of DNA. She suggests that its sugar-phosphate backbone is on the outside—an important clue that leads to the elucidation of DNA structure the following year.

U.S. biologists Alfred Day Hershey and Martha Chase use radioactive tracers to show that bacteriophages infect bacteria with DNA and not protein.

1953

April 25. English molecular biologist Francis Crick and U.S. biologist James Watson announce the discovery of the double helix structure of DNA, the basic material of heredity. They also theorize that if the strands are separated then each can form the template for the synthesis of an identical DNA molecule. It is perhaps the most important discovery in biology.

December 3. Scientists at the University of Iowa announce that they have induced human pregnancy using deep-frozen sperm, a process previously used to inseminate cattle.

British biochemists Archer Martin and A. T. James develop gas chromatography, a technique for separating the elements of a gaseous compound through differential absorption in a permeable solid.

English biochemist Frederick Sanger determines the structure of the insulin molecule. It is the largest protein molecule to have its chemical structure determined to date.

A model of the structure of deoxyribonucleic acid (DNA). (Corbis/Paul Seheult; Eye Ubiquitous)

U.S. biochemist Stanley Lloyd Miller shows that amino acids can be formed when simulated lightning is passed through containers of water, methane, ammonia, and hydrogen—conditions under which life may have arisen.

U.S. physiologists Eugene Aserinsky and Nathaniel Kleitman discover the rapid eye movements (REM) that characterize high brain activity during a period of sleep. Their discovery causes a revolution in the understanding of sleep processes, which had been thought to be quiet.

1954

February. U.S. medical doctor and microbiologist Jonas E. Salk, developer of the poliomyelitis vaccine, inoculates children against polio in Pittsburgh, Pennsylvania. This is the start of a program of mass inoculation.

Russian-born U.S. physicist George Gamow suggests that the genetic code consists of the order of nucleotide triplets in the DNA molecule.

U.S. scientists Gregory G. Pincus, Hudson Hoagland, and Min-Cheh Chang of the Worcester Foundation develop an oral contraceptive using the hormone norethisterone.

1955

April. Dr. Jonas E. Salk proclaims the success of his poliomyelitis vaccine, which has been tested in forty-four states.

English biochemist Dorothy Hodgkin elucidates the structure of vitamin B_{12}, a liver extract used in the treatment of pernicious anemia.

U.S. geneticists Joshua Lederberg and Norton Zinder discover that some viruses carry part of the chromosome of one bacterium to another; called transduction, the process becomes an important tool in genetics research.

1956

June. In a meeting of the American Medical Association, Dr. Jonas E. Salk and U.S. surgeon general Leonard A. Scheele predict that poliomyelitis will be eliminated in three years.

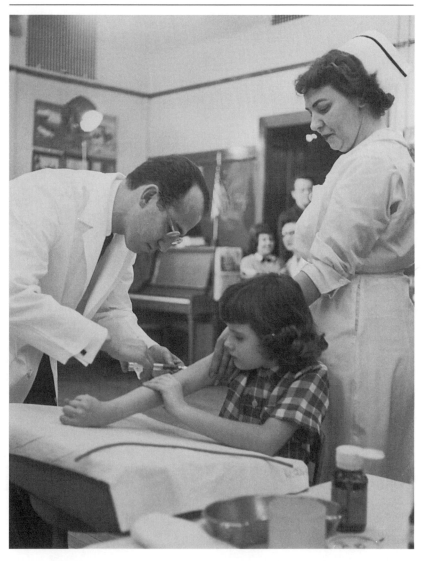

Dr. Jonas E. Salk injects Gail Rosenthal with the Salk polio vaccine during the 1954 field trials conducted in the Pittsburgh region. Nearly two million children, aged five to nine, took part in the field trials. (Corbis-Bettmann)

Romanian-born U.S. biologist George Palade discovers ribosomes, which contain RNA (ribonucleic acid).

Spanish-born U.S. molecular biologist Severo Ochoa discovers polynucleotide phosphorylase, the enzyme responsible for the synthesis of RNA (ribonucleic acid), which allows him to synthesize RNA.

U.S. biochemist Arthur Kornberg, using radioactively tagged nucleotides, discovers that the bacteria *Escherichia coli* use an enzyme, now known as DNA polymerase, to replicate DNA (deoxyribonucleic acid). It allows him to synthesize DNA in a test tube the following year.

U.S. biologists Maklon Hoagland and Paul Zamecnik discover transfer RNA (ribonucleic acid). This was later found to transfer amino acids, the building blocks of proteins, to the correct site on the messenger RNA.

1957

July. U.S. surgeon general Leroy E. Burney announces that scientists have established a link between cigarette smoking and lung cancer.

Russian-born U.S. engineer Vladimir Zworykin patents an instrument for observing microscopic organisms on a television screen.

Worcester Foundation scientist Gregory Pincus and Boston gynecologist John Rock begin a birth control pill trial in Puerto Rico.

1959

Austrian-born British biochemist Max Perutz determines the structure of hemoglobin.

1960

English anthropologist Jane Goodall discovers that chimpanzees can make tools, something only humans were thought capable of. She watches a chimpanzee fashion a blade of grass into a probe that can be poked into a termite mound to remove termites.

English biochemist John Kendrew, using X-ray diffraction techniques, elucidates the three-dimensional structure of the muscle protein myoglobin.

U.S. biochemist Robert Woodward and German biochemist Martin Strell independently synthesize chlorophyll.

1961

English molecular biologist Francis Crick and South African chemist Sydney Brenner discover that each base triplet on the DNA strand codes for a specific amino acid in a protein molecule.

French biochemists François Jacob and Jacques Monod discover messenger ribonucleic acid (mRNA), which transfers genetic information to the ribosomes, where proteins are synthesized.

1962

The Rockefeller and Ford Foundations found the Rice Research Institute in the Philippines and begin cross-breeding more than 10,000 different strains of rice.

In her book *Silent Spring*, U.S. biologist Rachel Carson draws attention to the dangers of chemical pesticides.

1963

Austrian zoologist Konrad Lorenz, in *On Aggression,* states that only humans intentionally kill their own species and that aggressive behavior is inborn but can be modified by the environment.

U.S. biochemist Robert Woodward synthesizes the plant chemical colchicine.

1964

The living brain of a rhesus monkey is isolated from its body by neurosurgeons at Cleveland General Hospital, Ohio.

U.S. divers spend nine days, at a depth of 58.5 m/192 ft, off the coast of Bermuda on board *Sealab* to study the effects of depth on the human mind and body.

U.S. zoologist William Hamilton recognizes the importance of altruistic behavior in animals, paving the way for the development of sociobiology.

1965

U.S. virologists Daniel Gajdusek and Clarence Gibbs transmit the diseases kuru and Creutzfeld-Jakob to primates.

U.S. biochemist Robert Woodward synthesizes the antibiotic cephalosporin C.

1966

English molecular biologist Francis Crick publishes *Of Molecules and Men*, in which he discusses the revolution occurring in molecular biology and its implications.

1967

U.S. biochemist Marshall Nirenberg establishes that mammals, amphibians, and bacteria all share a common genetic code.

U.S. scientist Charles Caskey and associates demonstrate that identical forms of messenger RNA produce the same amino acids in a variety of living beings, showing that the genetic code is common to all life forms.

1968

October. U.S. geneticists Mark Ptashne and Walter Gilbert separately identify the first repressor genes.

U.S. scientist Elso Sterrenberg Barghorn and associates report the discovery of the remains of amino acids in rocks three billion years old.

1969

February 15. English physiologist Robert Edwards of the Cambridge Physiological Laboratory in Cambridge, England, completes the first in-vitro fertilization of human egg cells.

September 15. The world's first heart and lung transplant is performed at the Stanford Medical Center in California.

U.S. geneticist Jonathan Beckwith and associates at the Harvard Medical School isolate a single gene for the first time.

1970

September. Indian-born U.S. biochemist Har Gobind Khorana assembles an artificial yeast gene from its chemical components.

U.S. biochemists Howard Temin and David Baltimore separately discover the enzyme reverse transcriptase, which allows some cancer viruses to transfer their RNA to the DNA of their hosts turning them cancerous—a reversal of the common pattern in which genetic information always passes from DNA to RNA.

U.S. geneticist Hamilton Smith discovers type II restriction enzyme that breaks the DNA strand at predictable places, making it an invaluable tool in recombinant DNA technology.

1971

January 6. Choh Hao Li and associates at the University of California Medical Center announce the synthesis of the human growth hormone somatotrophin.

A replacement knee joint is developed by Canadian surgeon Frank H. Gunston.

British physicians C. A. L. Bassett, R. J. Pawluk, and R. O. Becker discover that an electric current speeds up the healing of fractures.

English primatologist Jane Goodall publishes *In the Shadow of Man*, a study of chimpanzee behavior.

Polish-born U.S. endocrinologist Andrew Schally isolates the luteinizing hormone-releasing hormone (LH-RH), essential to human ovulation.

Surgeons develop the fiber-optic endoscope, making it possible to view inside the human body by inserting catheters into the arms or legs and manipulating them into organs, such as the heart.

The anticancer drug Taxol is isolated from the bark of the Pacific Yew tree.

The Royal College of Surgeons in England reports that deaths due to cigarette-smoking are comparable to the typhoid and cholera epidemics of the nineteenth century.

1972

English engineer Godfrey Hounsfield performs the first successful CAT (computerized axial tomography) scan, which provides cross-sectional X rays of the human body.

French mathematician René Frédéric Thom formulates catastrophe theory, an attempt to describe biological processes mathematically.

German researchers F. Eisenberger and C. Chaussy in Munich invent the lithotripter, a machine that disintegrates kidney stones using ultrasound; the device is commercialized in 1980.

Swiss researcher J.-F. Borel discovers the immunosuppressive properties of the drug cyclosporin-A.

The United States restricts the use of DDT because it is discovered that it thins the eggshells of predatory birds, lowering their reproductive rates.

U.S. microbiologist Daniel Nathans uses a restriction enzyme that splits DNA (deoxyribonucleic acid) molecules to produce a genetic map of the monkey virus (SV40), the simplest virus known to produce cancer; it is the first application of these enzymes to an understanding of the molecular basis of cancer.

U.S. paleontologists Stephen Jay Gould and Nils Eldridge propose the punctuated equilibrium model—the idea that evolution progresses in fits and starts rather than at a uniform rate.

Venezuelan-born U.S. immunologist Baruj Benacerraf and Hugh O'Neill McDevitt show immune response to be genetically determined.

1973

Representatives from eighty nations sign the Convention on International Trade in Endangered Species (CITES), which prohibits trade in 375 endangered species of plants and animals and the products derived from them, such as ivory; the United States does not sign.

The first calf is produced from a frozen embryo.

The U.S. firm 3M makes the first cochlear implant.

The U.S. Fish and Wildlife Bureau issues the first Endangered and Threatened Species List.

U.S. biochemists Stanley Cohen and Herbert Boyer develop the technique of recombinant DNA (deoxyribonucleic acid). Strands of DNA are cut by restriction enzymes from one species and then inserted into the DNA of another; this marks the beginning of genetic engineering.

1974

June. U.S. surgeon Jay Heimlich describes a method for saving people from choking on food. Quick upward thrusts of the heel of the hand placed in the person's abdomen dislodges the food. This becomes known as the "Heimlich maneuver."

July. The National Academy of Sciences advises a worldwide ban on recombinant DNA (deoxyribonucleic acid) experiments on the bacterium *Escherichia coli* for fear that a more virulent form may be created.

British-born Danish immunologist Niels Jerne proposes a network theory of the immune system.

The Parque Nacional da Amazonia is established in Brazil; with an area of 10,000 sq km/4,000 sq mi, it preserves a large area of tropical rain forest.

1975

After an absence of nearly one hundred years, Atlantic salmon return to spawn in the Connecticut River; the river was restocked in 1973 after efforts were made to end pollution.

In Cambridge, England, Argentine immunologist César Milstein and German immunologist Georges Köhler develop the first monoclonal antibodies—lymphocyte and myeloma tumor cell hybrids that are cloned to secrete unlimited amounts of specific antibodies.

In Oxford, England, British scientist Derek Brownhall produces the first clone of a rabbit.

Swiss scientists publish details of the first chemically directed synthesis of insulin.

The gel-transfer hybridization technique for the detection of specific DNA (deoxyribonucleic acid) sequences is developed; it is a key development in genetic engineering.

U.S. physiologist John Hughes discovers endorphins (morphinelike chemicals) in the brain.

1976

August 28. Indian-born U.S. biochemist Har Gobind Khorana and his colleagues announce the construction of the first artificial gene to function naturally when inserted into a bacterial cell, a major breakthrough in genetic engineering.

October 14. A committee to create standard regulations governing recombinant DNA (deoxyribonucleic acid) research is established by the International Council of Scientific Unions.

A mystery disease afflicts 182 people who attend the meeting of the American Legion in Philadelphia; 29 die within a month and the disease becomes known as Legionnaires' disease. It is caused by the bacterium *Legionella pneumophilia*.

Japanese molecular biologist Susumu Tonegawa demonstrates that antibodies are produced by large numbers of genes working in combination.

The first oncogene (cancer-inducing gene) is discovered by U.S. scientists Harold E. Varmus and J. Michael Bishop.

The first prenatal diagnosis using a gene-specific probe in amniotic fluid is performed.

U.S. biochemist Herbert Boyer and venture capitalist Robert Swanson found Genentech, the world's first genetic engineering company, in San Francisco.

U.S. scientists experiment with algae and microorganisms that consume crude oil as a means of clearing up oil spills.

1977

July 12. U.S. medical researcher Raymond Damadian produces the first images of human tissues using an NMR (nuclear magnetic resonance) scanner; used to detect cancer and other diseases without the need for X-rays, the scanner is based on the fact that electromagnetic fields cause some atomic nuclei to align themselves. The scanners become commercially available in the United States in 1984.

Dutch scientists discover that the wastes from incinerators are contaminated by dioxins—chemicals thought to cause cancer.

English biochemist Frederick Sanger achieves the first sequencing of an entire genome when he describes the full sequence of 5,386 bases in the DNA (deoxyribonucleic acid) of virus *phi*X174 in Cambridge, England.

German physician Andreas Gruentzig invents balloon angioplasty, a method of unclogging diseased arteries by guiding a catheter to the blocked area of the artery. A second catheter with a small balloon on the tip is passed through the first and then inflated to widen the artery for blood flow. The balloon is then deflated and removed.

Scientists discover chemosynthetically based animal communities around sulfurous thermal springs deep under the sea near the Galápagos Islands, Ecuador. These organisms use energy from chemical reactions rather than light energy, as in photosynthesis.

The United States signs the Convention on International Trade in Endangered Species (CITES).

English biochemists Frederick Sanger and Alan Coulson and U.S. molecular biologists Walter Gilbert and Allan Maxam develop a rapid gene-sequencing technique that uses gel electrophoresis, the diffusion of charged particles through a gel under the influence of an electric field.

U.S. scientist Herbert Boyer, of the firm Genentech, fuses a segment of human DNA (deoxyribonucleic acid) into the bacterium *Escherichia coli* which begins to produce the human protein somatostatin; this is the first commercially produced genetically engineered product.

1978

July 25. Louise Brown is born at Oldham Hospital in London, England; she is the first "test tube" baby. Having been unable to remove a blockage from her mother's Fallopian tube, gynecologist Patrick Steptoe and physiologist Robert Edwards removed an egg from her ovary, fertilized it with her husband's sperm, and reimplanted it in her uterus.

Cyclosporin A is introduced as an immunosuppressant drug in organ transplant surgery; it is given U.S. Food and Drug Administration approval in 1983.

The French surgeon Gabriel Coscas is the first to make medical use of an argon laser—in ophthamology.

1979

British obstetrician Ian Donald is the first to use ultrasound to examine a fetus.

H. Goodman and J. Baxter of the University of California, Berkeley, together with D. V. Goeddel of Genentech, announce the biosynthetic production of a human growth hormone.

Japanese researcher Ryochi Naito carries out the first experiment on a human being using artificial blood—he injects himself with the milky-looking fluid.

Researchers at Edinburgh University in Scotland successfully clone the hepatitis B viral antigen, opening the way for a successful vaccine.

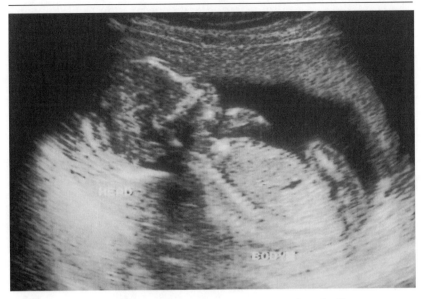

An ultrasound scan revealing the shape of a human fetus. (Corbis/Alan Towse; Ecoscene)

1980

The Swiss firm Biogen produces human interferon in bacteria for the treatment of viral diseases.

1981

A new family of deep-water stingray is named when a specimen of *Hexatrygon bickelli* is washed up on a South African beach.

Chinese scientists make the first clone of a fish (a golden carp).

Scientists at Ohio University inject genes from other species into the fertilized egg of a mouse, which develops into mice that carry the foreign gene in many of their cells. The gene is then passed on to their offspring, creating permanently altered (transgenic) animals; it is the first transfer of a gene from one animal species to another.

The genetic code for the hepatitis B surface antigen is discovered, creating the possibility of a bioengineered vaccine.

The Food and Drug Administration grants permission to Eli Lilly and Co. to market insulin produced by bacteria, the first genetically engineered product to go on sale.

U.S. geneticists Robert Weinberg, Geoffrey Cooper, and Michael Wigler discover that oncogenes (genes that cause cancer) are integrated into the genome of normal cells.

1982

Dolphins are discovered to possess magnetized tissues that aid in navigation; they are the first mammals discovered to have such tissues.

The naturally occurring chemical tribulin is discovered in the brain; it stimulates alertness.

The U.S. firm Applied Biosystems markets an automated gene sequencer that can sequence 18,000 DNA bases a day, compared with a few hundred a year by hand in the 1970s.

U.S. researcher Stanley Prusiner discovers prions (proteinaceous infectious particles); they are responsible for several neurological diseases including "mad cow disease" (first identified in 1986).

Using genetically engineered bacteria, the Swedish firm Kabivitrum manufactures human growth hormone.

1983

April. U.S. biochemist Kary Mullis invents the polymerase chain reaction (PCR), a method of multiplying genes or known sections of the DNA molecule a million times without the need for the living cell.

Geneticist James Gusella identifies the gene for Huntington's disease.

The skull of a creature called *Pakicetus* is discovered in Pakistan; estimated to be fifty million years old, it is intermediate in evolution between whales and land animals.

U.S. biologist Lynn Margulis discovers that cells with nuclei form by the synthesis of nonnucleated cells.

U.S. biologists Andrew Murray and Jack Szostak create the first artificial chromosome; it is grafted onto a yeast cell.

1984

April 14. The Texas Board of Education in the United States repeals a rule requiring evolution to be taught in schools as one of several possible theories of human origins, rather than as fact.

Allan Wilson and Russell Higuchi of the University of California, Berkeley, clone genes from an extinct animal, the quagga.

British geneticist Alec Jeffreys discovers that a core sequence of DNA (deoxyribonucleic acid) is almost unique to each person; this examination of DNA, known as "genetic fingerprinting," can be used in criminal investigations and to establish family relationships.

From studies of DNA (deoxyribonucleic acid), Charles Sibley and Jon Ahlquist argue that humans are more closely related to chimpanzees than to other great apes, differing in their DNA by only 1 percent, and that humans and apes diverged approximately five to six million years ago.

Robert Sinsheimer, the chancellor of the University of California, Santa Cruz, proposes that all human genes be mapped; the proposal eventually leads to the development of the Human Genome Project.

Sheep are successfully cloned by U.S. geneticist Steen A. Willadsen.

1985

French-born U.S. endocrinologist Roger Guillemin discovers inhibin, which suppresses follicle-stimulating hormone (FSH) secretion involved in male testicular function.

German researcher Bert Vallee and associates discover angiogenin, the factor that stimulates the growth of new blood vessels.

Researchers at the Massachusetts Eye and Ear Infirmary and the Whitehead Institute in Massachusetts isolate the first human cancer gene, retinoblastoma.

Researchers locate gene markers on chromosomes for cystic fibrosis and polycystic kidney disease.

U.S. zoologist Dian Fossey, who tried to protect endangered gorillas in Rwanda, is murdered. Poachers are suspected.

1987

April 24. Tests are carried out on gene-altered bacteria to aid agriculture, despite scientists' fears that loss of control is inevitable.

April. The U.S. Patent and Trademark Office announces its intention to allow the patenting of animals produced by genetic engineering.

October 10. The *New York Times* announces Dr. Helen Donis-Keller's mapping of all twenty-three pairs of human chromosomes, allowing the location of specific genes for the prevention and treatment of genetic disorders.

Chinese scientists insert genes controlling human growth hormones into goldfish and loach; they grow to four times the normal size.

Geneticist David C. Page and associates announce the discovery of a gene that initiates the development of male features in mammals.

Foxes in Belgium are immunized against rabies by using bait containing a genetically engineered vaccine, dropped from helicopters. The success of the experiment leads to a large-scale anti-rabies vaccination program.

German-born British geneticist Walter Bodmer and associates announce the discovery of a marker for a gene that causes cancer of the colon.

The development of the first bio-insecticides is announced in the United States; they eliminate insects without harming the environment.

The first genetically altered bacteria are released into the environment in the United States; they protect crops against frost.

The U.S. firm Diamond Sensor Systems develops the rapid blood analyzer; it can analyze the oxygen, carbon dioxide, pH, potassium, calcium, and hematocrit of the blood automatically in about two minutes.

The U.S. Supreme Court decrees invalid state laws requiring schools to give equal time to "creationist science," since they violate the Constitution's separation of church and state.

U.S. researcher Hari Reddi and his team at the National Institutes of Health discover bone morphogenetic protein—a protein that encourages bone growth. It is expected to speed up healing of fractures and bone damage done during surgery.

1988

Researchers at the Agency of Industrial Science and Technology of Japan develop a contractile synthetic material that contracts in acetone and expands in water, giving it the potential to be used in artificial limbs with muscular functions.

German-born U.S. biomedical researcher Rudolf Jaenisch and associates implant a human gene, connected with a hereditary disorder, into a mouse.

The Dutch firm CCA Biochem develops the biodegradable polymer polyactide; capable of being broken down by human metabolism, it is ideal for use in suturing threads, bone platelets, and artificial skin.

The first dairy cattle are produced by cloning embryos.

The Human Genome Organization (HUGO) is established in Washington, D.C.; scientists announce a project to compile a complete "map" of human genes.

U.S. researchers graft tissues from the bone marrow, spleen, thymus, and lymph nodes of human fetuses into mice lacking an immune system; the mice then develop an immune system identical to that of humans—a valuable tool in the development of vaccines.

1989

A lemur, *Allocebus tricholis*, previously thought to be extinct, is discovered in Madagascar.

Surgeons in Britain perform the first brain cell transplants, in which grafts of fetal brain tissue were used to treat Parkinson's disease.

Mary Pat Reeve studies a map of the human Y chromosome at the Whitehead Institute Genome Center in Boston, 17 February 1994. The Y chromosome is the only chromosome to have been fully mapped. (Corbis/Roger Ressmeyer)

Scientists in Britain introduce genetically engineered white blood cells into cancer patients, to attack tumors.

U.S. pediatric surgeon Michael R. Harrison and colleagues remove a fetus from its mother's womb, operate on its lungs, and return it to the womb.

U.S. scientists obtain the first visual image of a DNA (deoxyribonucleic acid) molecule.

1990

A four-year-old girl in the United States has the gene for adenosine deaminase inserted into her DNA (deoxyribonucleic acid); she is the first human to receive gene therapy.

French geneticist Pierre Chambon and associates announce the discovery of a gene that may be important in the development of breast cancer.

Six institutions are selected to participate in the project for mapping the genes of selected human chromosomes.

1991

British geneticists Peter Goodfellow and Robin Lovell-Badge discover the gene on the Y chromosome that determines sex.

The realization that all people with Down's syndrome will eventually develop early-onset Alzheimer's disease if they live long enough leads to the discovery that some forms of Alzheimer's disease are caused by a gene defect on chromosome 21.

1992

6 June. The United Nations Conference on Environment and Development is held in Rio de Janeiro, Brazil. It is attended by delegates from 178 countries, most of whom sign binding conventions to combat global warming and to preserve biodiversity (the latter convention was not signed by the United States).

An individual honey fungus *(Armallaria ostoyae)* is identified as the world's largest living thing. Discovered in Washington State and estimated to be between 500 and 1,000 years old, it has an underground network of hyphae covering 600 hectares/1,480 acres.

Epibatidine, a chemical extracted from the skin of an Ecuadorian frog, is identified as a member of an entirely new class of alkaloid. It is an organochlorine compound and is a powerful painkiller, about 200 times as effective as morphine.

Sperm cells are discovered to have odor receptors and may therefore reach eggs by detecting scent.

The U.S. biotechnology company Agracetus patents transgenic cotton, which has had a foreign gene added to it by genetic engineering.

U.S. biologist Philip Leder receives a patent for the first genetically engineered animal, the oncomouse, which is sensitive to carcinogens.

1993

Researchers isolate the first gene that predisposes individuals to cancer. About one in two hundred people carry the gene.

U.S. geneticist Dean Hammer and colleagues at the U.S. National Cancer Institute publish the approximate location of a gene that could predispose human males to homosexuality.

1994

May. The first genetically engineered food goes on sale in California and Chicago, Illinois. The "Flavr Savr" tomato is produced by the U.S. biotechnology company Calgene.

A new species of kangaroo is discovered in Papua New Guinea. Known locally as the bondegezou, it weighs 15 kg/7 lb and is 1.2 m/3.9 ft in height.

Short stretches of dinosaur DNA are extracted from unfossilized bone retrieved from coal deposits approximately eighty million years old.

1995

July. The U.S. government approves experimentation of genetically altered animal organs as transplants into humans.

A genetically engineered potato is developed that contains the gene for Bt toxin, a natural pesticide produced by a soil bacterium. The potato plant produces Bt within its leaves.

Australian geneticists produce a genetically engineered variety of cotton that contains a gene from a soil bacteria that kills the cotton bollworm and native budworm.

U.S. scientists successfully germinate bacterial spores extracted from the gut of a bee fossilized in amber forty million years ago.

1996

January 18. New Zealand ornithologist Gavin Hunt reveals that crows on the island of New Caledonia in the South Pacific make tools out of leaves and twigs which they use to reach insects in dead wood—something only chimpanzees and humans were thought capable of.

February 21. U.S. researcher Joanna Fowler and her colleagues at the Brookhaven National Laboratory in Upton, New York, report that smoking reduces the enzyme monamine oxidase B in the brain, which leads to an increase in the amount of dopamine, a chemical that helps regulate mood, movement, and the reinforcement of behavior patterns. It is the first identification of a mechanism of cigarette addiction. High levels of dopamine are also found in other addictive drugs.

April 3. The World Health Organization (WHO) announces the development, by an international team of scientists, of a weekly injection that reduces sperm production to a negligible level, while leaving sexual performance unimpaired.

May 9. Scientists at the National Institute of Allergy and Infectious Disease discover a protein, "fusin," which allows the HIV virus to fuse with a human immune system cell's outer membrane and inject genetic material. Its presence is necessary for the AIDS virus to enter the cell.

August. U.S. geneticists clone two rhesus monkeys from embryo cells.

October 3. British scientists announce meteorite evidence of life on Mars, supporting claims made by NASA and U.S. scientists in August.

October 4. The World Conservation Union (IUCN) publishes the latest Red List of endangered species, which notes more than one thousand mammals, far more than on previous lists. The organization believes it has underestimated the risks on habitats from pollution and that the number of endangered species is greater than previously thought.

October 17. U.S. researchers from the University of Texas and the Beckman Research Institute based in Hope, California, announce the discovery that cigarette smoke alters a gene that suppresses the uncontrolled growth of cells that cause tumors. It is the first direct evidence for the statistical link between cigarette smoking and cancer.

The shrimp *Synalpheus regalis*, which lives in sponges in the coral reefs of Belize, is discovered to live in social colonies like those of ants. All the shrimps are the offspring of a single reproductive female, care of young is cooperative, and larger individuals act to defend the colony. This type of social organization, called eusociality, was previously known to occur only among some types of insects and among naked mole rats.

Two U.S. dentists discover a new muscle running from the jaw to just behind the eye socket. Measuring about 3 cm/1 in, it helps to support and raise the jaw.

1997

January 31. An international team of scientists reports the discovery of the gene for a type of glaucoma (open angle glaucoma, one of the commonest forms of blindness) that threatens up to one in fifty people in later life. The gene, on human chromosome number 1, codes for a protein called TIGR, which could be linked to the destruction of nerve cells in the eye.

February 20. U.S. researchers Jin Meng and André R. Wyss describe impressions and casts of fossilized hair recovered from sixty-million-year-old rocks in Inner Mongolia, China. The casts show that hair is an extremely ancient feature of mammals, going right back to their origin.

February 27. Scottish researcher Ian Wilmut of the Roslin Institute in Edinburgh, Scotland, announces that British geneticists have cloned an adult sheep. A cell was taken from the udder of the mother sheep and its DNA (deoxyribonucleic acid) combined with an unfertilized egg that had had its DNA removed. The fused cells were grown in the laboratory and then implanted into the uterus of a surrogate mother sheep. The resulting lamb, Dolly, came from an animal that was six years old. This is the first time cloning has been achieved using cells other than reproductive cells. The news is met with international calls to prevent the cloning of humans.

February. U.S. genetic scientist Don Wolf announces the production of monkeys cloned from embryos. It is a step closer to cloning humans and raises acute philosophical issues.

February. U.S. zoologists Bill Detrich and Kirk Malloy show that the increased ultraviolet radiation caused by the hole in the ozone layer above Antarctica kills large numbers of fish in the Southern Ocean. Because their transparent eggs and larvae stay near the surface for up to a year, they are exposed to the full force of the ultraviolet rays. It is the first time ozone depletion in the Antarctic has been shown to harm organisms larger than one-celled marine plants.

Dolly, the famous cloned sheep, in her stall. (Reuters/Handout/Archive Photos)

March. President Bill Clinton announces a ban on using federal funds to support human cloning research and calls for a moratorium on this type of scientific investigation. He also asks the National Bioethics Advisory Commission to review and issue a report on the ramifications that cloning would have on humans.

April 30. U.S. researchers at the University of Pennsylvania Medical Center report that chimpanzees immunized with a new kind of DNA-based vaccine and then infected with 250 times the HIV needed to cause infection had no trace of the virus after forty-eight weeks. The report again raises hopes for a vaccine.

May 8. U.S. AIDS researcher David D. Ho and colleagues show how aggressive treatment of HIV-1 infection with a cocktail of three antiviral drugs can drive the virus to below the limits of conventional clinical detection within eight weeks.

May 16. U.S. geneticists identify a clock gene in chromosome 5 in mice that regulates the circadian rhythm.

May 17–May 20. At a meeting of the American Society of Clinical Oncology, researchers announce the development of vaccines that cause the immune system to shrink certain cancers, such as those attacking the skin, breast, prostate, and ovaries. Unlike the normal preventive vaccines, the new vaccines fight tumors that already exist. They use components of the cancer to provoke white blood cells to attack the invader.

June 3. U.S. geneticist Huntington F. Wilard constructs the first artificial human chromosome. He inserts telomeres (which consist of DNA [deoxyribonucleic acid] and protein on the tips of chromosomes) and centromeres (specialized regions of DNA within a chromosome) removed from white blood cells into human cancer cells, which are then assembled into chromosomes about one-tenth the size of normal chromosomes. The artificial chromosome is successfully passed on to all daughter cells.

June 11. English behavioral scientist David Skuse claims that boys and girls differ genetically in the way they acquire social skills. Girls acquire social skills intuitively and are "preprogrammed," while boys have to be taught. This has important implications for education.

June 26. The second Earth Summit takes place in New York City. Delegates report on progress since the 1992 Rio Summit and note that progress on the Rio biodiversity convention has been slower than on the convention on climate. The delegates fail to agree on a deal to address the world's escalating environmental crisis. Dramatic decreases in aid to the so-called Third World countries, which the 1992 summit promised to increase, are at the heart of the breakdown.

June 26. U.K. physiologist Stephen O'Rahilly and colleagues show that human obesity can be caused by a mutation in the gene that produces the hormone leptin. The brain uses leptin as a measure of the amount of body fat; an increase in fat means an increase in leptin, which suppresses the appetite and increases the burning of excess calories. A deficiency of leptin, caused by a malfunction in its gene, means that the brain receives no signal of body fat levels, and hence fails to trigger increased metabolism. Pharmaceutical companies compete for the rights to develop dieting products based on this knowledge.

July 11. Teams of researchers from Germany and the United States use mitochondrial DNA (deoxyribonucleic acid) extracted from the original fossils of Neanderthal Man, discovered in the Neander Valley near Düsseldorf, Germany, in 1856, to confirm that Neanderthal Man and modern humans diverged evolutionarily about six hundred thousand years ago. The researchers' work supports the theory that modern humans arose recently in Africa as a distinct species and replaced Neanderthals with little or no interbreeding, while the Neanderthals became extinct without evolving into modern humans.

July 18. Japanese scientist Yoshinori Kuwabara announces that his team has successfully grown goat embryos in an acrylic tank. The embryos, removed from their mother at seventeen weeks into pregnancy, are placed in the tank filled with a liquid that simulates amniotic fluid. The placenta is replaced by a machine that pumps oxygen and nutrients into the embryo's blood. At twenty weeks' gestation the goat is "born." At present the procedure can only be done late in fetal development.

August 4. U.S. researchers Sidney Altman, Cecilia Guerreir-Takada, and Reza Salavati discover a gene-transfer method of disabling the genes in disease organisms that allow them to neutralize common antibiotics. This makes the bacteria vulnerable, once again, to treatment with antibiotics and combats the growing problem of bacterial drug resistance. The biomedical company Innovir Laboratories works on developing the process to combat viruses such as those responsible for hepatitis B and C.

August 7. Canadian researcher Suzanne W. Simard and colleagues announce the discovery that trees use the threadlike growths of fungi called mycorrhizae, which infest their roots and connect the trees together underground to exchange food resources with other trees. This finding suggests that forest trees succeed as cooperative communities rather than competing individuals.

August 22. Scientists from the World Wide Fund for Nature (WWF) announce the discovery of a new species of muntjac deer in Vietnam. A dwarf species weighing only about 16 kg/35 lb, it has antlers the length of a thumbnail and lives at altitudes of 457–914 m/1,500–3,000 ft.

August. U.S. geneticist Craig Venter and colleagues publish the genome of the bacterium *Helicobacter pylori,* a bacterium that infects half the world's population and is the leading cause of stomach ulcers. It is the sixth bacterium to have its genome published, but is clinically the most important. It has 1,603 putative genes, encoded in a single circular chromosome that is 1,667,867 nucleotide base-pairs of DNA (deoxyribonucleic acid) long. Complete genomes are being published with increasing frequency as gene-sequencing techniques improve.

September 7. Australian researcher William de la Mare, using old whaling records that record data on every whale caught since the 1930s, including the ship's latitude, announces the discovery that Antarctic sea ice could have decreased by up to 25 percent between the mid-1950s and the 1970s. The finding has major implications, both for global climate conditions as well as for whaling.

September 18. U.S. geneticist Bert Vogelstein and colleagues demonstrate that the p53 gene, which is activated by the presence of carcinogens, induces cells to commit "suicide" by stimulating them to produce large quantities of poisonous chemicals, called "reactive oxygen species" (ROS). The cells literally poison themselves. It is perhaps the human body's most effective way of combating cancer. Many cancers consist of cells with a malfunctioning p53 gene.

October 2. U.K. scientists Moira Bruce and, independently, John Collinge and their colleagues show that the new variant form of the brain-wasting Creutzfeldt-Jakob disease (CJD) is the same disease as bovine spongiform encephalopathy (BSE or "mad cow disease") in cows.

U.S. microbiologists bring to life a previously unknown *Staphylococcus* bacterium species from spores preserved in amber.

U.S. scientists at the National Human Genome Research Institute announce the discovery of a gene that causes Parkinson's disease. The gene produces a protein called alpha synuclein. When the instructions of the gene go wrong, the protein's structure is affected and this causes the buildup of deposits on brain cells that is usually seen in Parkinson's sufferers.

1998

There are 3,214 species of rare and endangered plants in the United States.

January 7. Doctors meeting at the World Medical Association's conference in Hamburg, Germany, call for a worldwide ban on human cloning. President Bill Clinton calls for legislation banning cloning the following day.

January 30. U.S. scientist Angela Christiano of Columbia University, New York, publishes a study that identifies a "hairless" gene that causes severe hair loss.

February 20. Researchers at the University of Texas, in conjunction with the British company SmithKline Beecham, publish a study in which they identify a hormone that triggers hunger in humans. Scientists hope the discovery will lead to potential treatments of appetite disorders.

March 2. U.S. scientist Dennis McFadden and his team of researchers from the University of Texas in Austin report a physiological difference between lesbians and heterosexual women that could influence sexual orientation. Echoes produced in the ears of lesbians in response to clicking sounds were weaker than those of their heterosexual counterparts.

April 22. Scientists at the Public Health Laboratory Service in London, England, report the discovery of a bacterium, *Pseudonas aeruginosa,* that is resistant to all known antibiotics. It causes a wide range of infections in people with impaired immune systems.

July. Japanese researchers clone twin calves from adult cells. This is the first incidence of a successful replication of the cloning method used to produce Dolly the sheep by British researchers in 1997.

October. Researchers find that people create new brain cells during adulthood, a discovery that suggests that someday the brain might be able to repair damage from disease and injury.

November. Researchers discover fossil evidence of the world's oldest flowering plant: a 142-million-year-old twiglike plant with peapod-shaped fruit. Researchers say the fossilized plant represents an early evolutionary stage when plants were developing the system that later evolved into the power to produce fruit, grain, and brilliantly colored and fragrant flowers.

1999

March 15. Scientists working on the Human Genome Project announce that the first working draft of the human genome will be completed February 2000, a year ahead of schedule. The draft will include 90 percent of the three billion human DNA sequences and will lay the groundwork for the completion of the final sequence by 2003.

3

Biographical Sketches

Berg, Paul (1926–)

U.S. molecular biologist who spliced the DNA from an animal tumor virus (SV40) and the DNA from a bacterial virus and combined it into a single hybrid, using gene-splicing techniques (developed by others) in 1972. For his work on recombinant DNA, he shared the 1980 Nobel Prize for Chemistry.

Berg was born in New York and educated at Pennsylvania State University and Case Western Reserve University. Between 1955 and 1974 he held several positions at Washington University.

In 1956 Berg identified an RNA molecule (later known as a transfer RNA) that is specific to the amino acid methionine. He then perfected a method for making bacteria accept genes from other bacteria. This genetic engineering can be extremely useful for creating strains of bacteria to manufacture specific substances, such as interferon. But there are also dangers: a new, highly virulent pathogenic microorganism might accidentally be created, for example. Berg has therefore advocated restrictions on genetic engineering research.

Calvin, Melvin (1911–1997)

U.S. chemist who, using radioactive carbon-14 as a tracer, determined the biochemical processes of photosynthesis, in which green plants use chlorophyll to convert carbon dioxide and water into sugar and oxygen. He was awarded the Nobel Prize for Chemistry in 1961.

Calvin was born in St Paul, Minnesota, and studied at Michigan College of Mining and Technology and the University of Minnesota. From 1937 he was on the staff of the University of California, becoming professor in 1947.

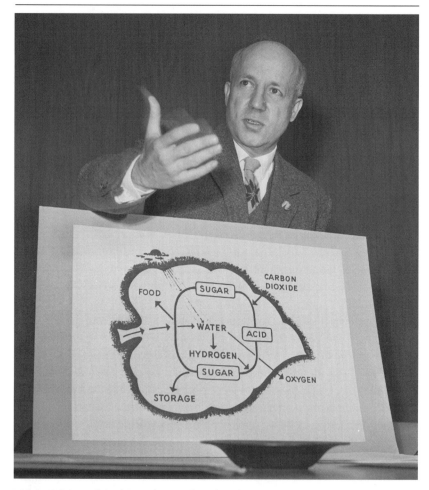

Melvin Calvin explains his work on carbon dioxide assimilation in plants at a press conference in California, 3 November 1961. (Corbis-Bettmann)

Calvin began work on photosynthesis in 1949, studying how carbon dioxide and water combine to form carbohydrates such as sugar and starch in a single-celled green alga, *Chlorella*. He showed that there is in fact a cycle of reactions (now called the Calvin cycle) involving an enzyme as a catalyst.

Carson, Rachel Louise (1907–1964)

U.S. biologist, writer, and conservationist whose book *Silent Spring*, published in 1962, attacked the indiscriminate use of pesticides, especially DDT, and inspired the creation of the modern environmental movement.

Carson was born in Springdale, Philadelphia, and educated at Pennsylvania College for Women and Johns Hopkins University. She worked first at the University of Maryland and the Woods Hole Marine Biological Laboratory in Massachusetts, and then as an aquatic biologist with the U.S. Fish and Wildlife Service from 1936 to 1949, becoming its editor-in-chief until 1952.

Undated photograph of Rachel Carson. (Corbis/Underwood & Underwood)

Her first book, *The Sea Around Us* (1951), was a best-seller and won several literary awards. It was followed by *The Edge of the Sea* (1955), an ecological exploration of the seashore. While writing about broad scientific issues of pollution and ecological exploitation, she also raised important issues about the reckless squandering of natural resources by an industrial world.

Chase, Mary Agnes Meara (1869–1963)

U.S. botanist and suffragist who made an outstanding contribution to the study of grasses. During the course of several research expeditions she collected many plants previously unknown to science, and her work provided much information about naturally occurring cereals and other food crops.

Meara was born in Iroquois County, Illinois, and was self-educated. From 1903 she worked in Washington, D.C., for the U.S. Department of Agriculture Bureau of Plant Industry and Exploration, and became the principal scientist for agrostology (study of grasses). She was politically active in various reform movements, especially those for female suffrage, and on this account was jailed and forcibly fed during World War I.

During the course of several research expeditions she collected many plants previously unknown to science, and her work provided much information about naturally occurring cereals and other food crops. Chase was responsible for work in modernizing and extending the national grass herbarium. She traveled widely, collecting plants from several regions of North and South America, and also visiting European research institutes and herbaria during the 1920s. Altogether she collected more than twelve thousand plants for the herbarium.

Crick, Francis Harry Compton (1916–)

When the war finally came to an end, I was at a loss as to what to do. . . . Only gradually did I realize that [my] lack of qualification could be an advantage. By the time most scientists have reached age thirty they are trapped by their own expertise. They have invested so much effort in one particular field that it is often extremely difficult . . . to make a radical change. . . . Since I essentially knew nothing, I had an almost completely free choice. . . .

—*Francis Crick*

English molecular biologist who researched the molecular structure of DNA and the means whereby characteristics are transmitted from one generation to another. For this work he was awarded a Nobel Prize for Physiology or Medicine (with Maurice Wilkins and James Watson) in 1962.

Crick was born in Northampton and studied physics at University College, London. During World War II he worked on the development of radar. He then went to do biological research at Cambridge. In 1977 he became a professor at the Salk Institute in San Diego, California.

Using Wilkins's and others' discoveries, Crick and Watson put forward the idea that DNA consists of a double helix made up of two parallel chains of alternate sugar and phosphate groups linked by pairs of organic bases. They built molecular models that also explained how genetic information could be coded in the sequence of organic bases. Crick and Watson published their work on the proposed structure of DNA in 1953. Their model is now generally accepted as correct.

Later, this time working with South African Sidney Brenner, Crick demonstrated that each group of three adjacent bases (he called a set of three bases a codon) on a single DNA strand codes for one specific amino acid. He also helped to determine codons that code for each of the twenty main amino acids. Furthermore, he formulated the adaptor hypothesis, according to which adaptor molecules mediate between messenger RNA and amino acids. These adaptor molecules are now known as transfer RNAs.

Dobzhansky, Theodosius (1900–1975)

Ukrainian-born U.S. geneticist who established evolutionary genetics as an independent discipline. He showed that genetic variability between individuals of the same species is very high and that this diversity is vital to the process of evolution.

Dobzhansky was born Feodosy Grigorevich Dobrzhansky in Nemirov and studied at Kiev. After teaching at Kiev and Leningrad Universities, he emigrated to the United States 1927. He was at the California Institute of Technology from 1929 to 1940. He became professor of zoology at Columbia University in New York City in 1940; worked at the Rockefeller Institute (later Rockefeller University) from 1962 to 1971; then he moved to the University of California at Davis.

Dobzhansky's book *Genetics and the Origin of Species* (1937) was the first significant synthesis of Darwinian evolutionary theory and

Mendelian genetics. Dobzhansky also proved that there is a period when speciation is only partly complete and during which several races coexist.

His book *Mankind Evolving* (1962) had great influence among anthropologists and he wrote again on the philosophical aspects of evolution in *The Biological Basis of Human Freedom* (1956) and *The Biology of Ultimate Concern* (1967).

Elion, Gertrude Belle (1918–)

> *Stop being polite. Challenge people. Be prepared to ask the impolite questions.*
>
> —*Gertrude Belle Elion*

U.S. biochemist who shared the Nobel Prize for Physiology or Medicine in 1988 with her colleague George Herbert Hitchings and British physiologist James Black for her work on the development of drugs to treat cancer, gout, malaria, and various viral infections.

Elion was born in New York City and graduated from Hunter College in 1937 before obtaining her M.S. from New York University. In 1944 she joined the laboratories of Burroughs-Wellcome in North Carolina. She continued this association for more than four decades, being named scientist emerita at Burroughs-Wellcome in 1983. In 1973 she became adjunct professor of pharmacology at the University of North Carolina at Chapel Hill and became research professor of pharmacology at Duke University in 1983. She has continued in her positions at Duke and Chapel Hill and has been involved in various capacities with the World Health Organization.

Elion is best known for her work in developing new drugs, particularly those that inhibit the synthesis of DNA by diseased cells. With her colleague Hitchings, she investigated the chemistry and function of two important components of DNA, pyrimidine and purine. From this research, they developed new drugs to treat leukemia, malaria, kidney stones, gout, herpes, and AIDS. Elion also helped to develop early forms of immunosuppressive drugs to enable transplant patients to tolerate the presence of tissues from an unrelated donor.

Evans, Alice Catherine (1881–1975)

U.S. microbiologist whose research into the bacterial contamination of milk led to the recognition of the danger of unpasteurized milk. Brucellosis in humans and cattle had been thought to be two separate diseases until Evans published her findings in 1918. As a result of her research the incidence of brucellosis was greatly reduced when the dairy industry finally accepted that all milk should be pasteurized, in the 1930s.

Evans was born in Neath, Pennsylvania, and studied at Cornell University and the University of Wisconsin, after which she took a research post at the U.S. Department of Agriculture where she studied the bacteriology of milk and cheese. In 1918 she moved to the Hygienic Laboratories of the United States Public Health Service to research into epidemic meningitis and influenza as well as milk flora.

Fossey, Dian (1938–1985)

U.S. zoologist who was almost completely untrained when she was sent by Louis Leakey into the African wild. From 1975, she studied mountain gorillas in Rwanda and discovered, by living in close proximity to them, that they led generally peaceful family lives, though there were some incidences of infanticide. She also observed that females transferred to nearby established groups. Fossey was murdered by poachers whose snares she had cut during her relentless crusade against poaching within the Parc National des Volcans in Rwanda.

Frisch, Karl von (1886–1982)

Austrian zoologist who was cofounder (with fellow Austrian Konrad Lorenz) of ethology, the study of animal behavior. Born in Vienna, Frisch studied at universities in Munich and Trieste and taught in Munich. He specialized in bees, discovering how they communicate the location of sources of nectar by movements called "dances." He was awarded the Nobel Prize for Physiology or Medicine 1973 together with Lorenz and Dutch-born British zoologist Nikolaas Tinbergen for their pioneering work in ethology.

Gallo, Robert Charles (1937–)

U.S. scientist who is credited with identifying the virus responsible for AIDS. Gallo discovered the virus, now known as human immunodeficiency virus (HIV), in 1984; the French scientist Luc Montagnier of the Pasteur Institute, Paris, discovered the virus, independently, in 1983. The sample in which Gallo discovered the virus was supplied by Montagnier, and it has been alleged that this may have been contaminated by

Robert Gallo discusses the causes of AIDS at a press conference in Washington, 23 April 1984. (Corbis-Bettmann)

specimens of the virus isolated by Montagnier a few months earlier. Gallo spent thirty years at the National Cancer Institute, Maryland. In 1996 he headed the newly founded Institute of Virology in Baltimore.

Goodall, Jane (1934–)

I want to make [people] aware that animals have their own needs, emotions, and feelings—they matter. . . . I want to give kids a passion, an understanding and awareness of the wonder of animals.

—*Jane Goodall*

English primatologist and conservationist who has studied the chimpanzee community on Lake Tanganyika since 1960 and is a world authority on wild chimpanzees.

Goodall was born in London. She left school at eighteen and worked as a secretary and a film production assistant, until she had an opportunity to work for anthropologist Louis Leakey in Africa. She began to study the chimpanzees at the Gombe Stream Game Reserve, on Lake Tanganyika. Goodall observed the lifestyles of chimpanzees in their natural habitats,

Jane Goodall encounters a curious chimp at the Gombe Stream Research Center in Africa, 23 January 1972. (Corbis-Bettmann)

discovering that they are omnivores, not herbivores as originally thought, and that they have highly developed and elaborate forms of social behavior. She obtained a Ph.D. from Cambridge University in 1965, despite the fact that she had never been an undergraduate. Her books include *In the Shadow of Man* (1971), *The Chimpanzees of Gombe: Patterns of Behavior* (1986), and *Through a Window* (1990). In the 1990s, most of Goodall's time has been devoted to establishing sanctuaries of illegally captured chimpanzees, fund-raising, and speaking out against the unnecessary use of animals in research.

Gould, Stephen Jay (1941–)

U.S. paleontologist and writer. In 1972 he proposed the theory of punctuated equilibrium, suggesting that the evolution of species did not occur at a steady rate but could suddenly accelerate, with rapid change occurring over a few hundred thousand years. His books include *Ever Since Darwin* (1977), *The Panda's Thumb* (1980), *The Flamingo's Smile* (1985), and *Wonderful Life* (1990).

Gould was born in New York and studied at Antioch College, Ohio, and Columbia University. He became professor of geology at Harvard in 1973 and was later also given posts in the departments of zoology and the history of science.

Gould has written extensively on several aspects of evolutionary science, in both professional and popular books. His *Ontogeny and Phylogeny* (1977) provided a detailed scholarly analysis of his work on the developmental process of recapitulation. In *Wonderful Life,* he drew attention to the diversity of the fossil finds in the Burgess Shale Site in Yoho National Park, Canada, which he interprets as evidence of parallel early evolutionary trends extinguished by chance rather than natural selection.

Harvey, Ethel Browne (1885–1965)

U.S. embryologist and cell biologist who discovered the mechanisms of cell division, using sea-urchin eggs as her experimental model.

Ethel Browne Harvey was born in Baltimore and studied at Columbia University, New York. After her marriage, she worked part-time and visited several marine laboratories; from 1931 she was an independent research worker attached to the biology department of Princeton. She was a frequent visitor to the Stazione Zoologica in Naples.

Harvey's work concentrated on the role in cell fertilization and development of nonnuclear cell components in the cytoplasm. She undertook morphological studies and physiological experiments to examine the factors that affect the process of cell division and was able to stimulate division in fragments of sea-urchin eggs that contained no nucleus. This was an important contribution to unraveling the connections between different cellular structures in controlling cell division and development.

Hyman, Libbie Henrietta (1888–1969)

U.S. zoologist whose six-volume *The Invertebrates* (1940–1968) provided an encyclopedic account of most phyla of invertebrates. Initially she worked on flatworms, but soon extended her investigations to a wide spread of invertebrates, especially their taxonomy (classification) and anatomy.

Hyman was born in Des Moines, Iowa, and studied at the University of Chicago, where she remained as a research assistant until 1930. She then traveled to several European laboratories, working for a period at the Stazione Zoologica in Naples before returning to New York City to begin writing *The Invertebrates,* for which she was given office and laboratory space, but no salary, by the American Museum of Natural History.

Jacob, François (1920–)

French biochemist who, with Jacques Monod and André Lwoff, pioneered research into molecular genetics and showed how the production of proteins from DNA is controlled. They shared the Nobel Prize for Physiology or Medicine in 1965.

Jacob was born in Nancy and studied at the University of Paris. In 1950 he joined the Pasteur Institute in Paris as a research assistant, becoming head of the Department of Cellular Genetics 1964 and also professor of cellular genetics at the Collège de France.

Jacob began his work on the control of gene action in 1958, working with Lwoff and Monod. It was known that the types of proteins produced in an organism are controlled by DNA, and Jacob focused his research on how the amount of protein is controlled. He performed a series of experiments in which he cultured the bacterium *Escherichia coli* in various media to discover the effect of the medium on enzyme production. He and his team found that there were three types of gene concerned with the production of each specific protein.

Khorana, Har Gobind (1922–)

I do have a basic faith that survival of our civilization is not even going to
be possible without proper use of science.

—*Har Gobind Khorana*

Indian-born U.S. biochemist who in 1976 led the team that first synthesized a biologically active gene. In 1968 he shared the Nobel Prize for Physiology or Medicine for research on the chemistry of the genetic code and its function in protein synthesis. Khorana's work provides much of the basis for gene therapy and biotechnology.

Khorana was born in Raipur in the Punjab, now in Pakistan. He studied at Punjab University, in the United Kingdom at Liverpool, and in Switzerland at Zürich, returning to Britain 1950 to work at Cambridge. He has held academic posts in the United States and Canada, becoming professor at the University of Wisconsin in 1962 and at the Massachusetts Institute of Technology in 1970.

Khorana systematically synthesized every possible combination of the genetic signals from the four nucleotides known to be involved in determining the genetic code. He showed that a pattern of three nucleotides, called a triplet, specifies a particular amino acid (the building blocks of proteins). He further discovered that some of the triplets provided punctuation marks in the code, marking the beginning and end points of protein synthesis.

Krebs, Edwin Gerhard (1918–)

U.S. physician who won the Nobel Prize for Physiology or Medicine with U.S. scientist Edmond Fischer in 1992 for his discovery of protein phosphorylation (the chemical bonding of a phosphate molecule to a protein) as a control mechanism in the metabolic activity of mammalian cells.

Krebs was born in Lansing, Iowa, and graduated from the University of Illinois. He later joined the staff of the medical school of the University of Washington, where in the 1950s he began working with Fischer on the fundamental chemical reactions that regulate cell metabolism.

Krebs and Fischer characterized a group of enzymes, called protein kinases, which change proteins from their inactive to active form by triggering the chemical bonding of a phosphate group to the protein. This phosphorylation is the underlying switch that starts and stops a variety of cell functions, from breakdown of fats to the generation of chemical energy in response to hormonal and other signals. They determined adenosine triphosphate (ATP), a nucleotide molecule found in all cells, to be the energy-transporting compound that donated the phosphate group. Working on muscle tissue, they also showed that protein phosphorylation was the underlying mechanism that accounted for the reversible modification of glycogen phosphorylase.

Hundreds of protein kinases have been found since their discovery and it has been estimated that perhaps 1 percent of genes encode one sort of protein kinase or another. Indeed it is now evident that phosphorylation controls virtually every important reaction in cells and provides the basis for understanding how integrated cellular behavior is regulated by both intracellular control mechanisms and extracellular signals.

Krebs, Hans Adolf (1900–1981)

German-born British biochemist who discovered the citric acid cycle, also known as the Krebs cycle, the final pathway by which food molecules are converted into energy in living tissues. For this work he shared the Nobel Prize for Physiology or Medicine in 1953.

Krebs was born in Hildesheim and studied at the universities of Göttingen, Freiburg, Munich, Berlin, and Hamburg. In 1933, with the rise to power of the Nazis, he moved to the United Kingdom, initially to Cambridge, then to Sheffield in 1935. He was professor at Sheffield University from 1945 to 1954, and at Oxford from 1954 to 1967.

Krebs first became interested in the process by which the body degrades amino acids. He discovered that nitrogen atoms are the first to be removed (deamination) and are then excreted as urea in the urine. He then investigated the processes involved in the production of urea from the removed nitrogen atoms, and by 1932 he had worked out the basic steps in the urea cycle.

Lederberg, Joshua (1925–)

U.S. geneticist who showed that bacteria can reproduce sexually, combining genetic material so that offspring possess characteristics of both parent organisms. In 1958 he shared the Nobel Prize for Physiology or Medicine with George Beadle and Edward Tatum.

Lederberg was born in New Jersey and studied at Columbia and at Yale, where he worked with Tatum. He was at the University of Wisconsin from 1947 to 1959, rising to professor, and moved in 1959 to Stanford

American geneticist and biologist Joshua Lederberg, 1958. (Corbis/Hulton-Deutsch Collection)

University in California, becoming director of the Kennedy Laboratories of Molecular Medicine in 1962.

Lederberg is a pioneer of genetic engineering, a science that relies on the possibility of artificially shuffling genes from cell to cell. He realized 1952 that bacteriophages, viruses that invade bacteria, can transfer genes from one bacterium to another, a discovery that led to the deliberate insertion by scientists of foreign genes into bacterial cells.

Lorenz, Konrad Zacharias (1903–1989)

Regarding culture as a living system and considering its disturbances in the light of illnesses led me to the opinion that the main threat to humanity's further existence lies in that which may well be called mass neurosis. One might also say that the main problems with which humanity is faced are moral and ethical problems.

—Konrad Lorenz

Austrian ethologist who studied the relationship between instinct and behavior, particularly in birds, and described the phenomenon of imprinting in 1935. His books include *King Solomon's Ring* (1952), on animal behavior, and *On Aggression* (1966), which explores human behavior. In 1973 he shared the Nobel Prize for Physiology or Medicine with Dutch-born British zoologist Nikolaas Tinbergen and Austrian zoologist Karl von Frisch.

Lorenz was born in Vienna and studied medicine there and in the United States at Columbia University in New York City. In 1940 he was appointed professor of general psychology at the Albertus University in Königsberg, Germany. Lorenz sympathized with Nazi views on eugenics and in 1938 applied to join the Nazi party. From 1942 to 1944 he was a physician in the German army and then spent four years in the Soviet Union as a prisoner of war. Returning to Austria, he successively headed various research institutes.

Together, Lorenz and Tinbergen discovered how birds of prey are recognized by other birds: all birds of prey have short necks, and the sight of any bird—or even a dummy bird—with a short neck causes other birds to fly away.

Luria, Salvador Edward (1912–1991)

Italian-born U.S. physician who was a pioneer in molecular biology, especially the genetic structure of viruses. Luria was a pacifist and was identified with efforts to keep science humanistic. He shared the Nobel Prize for Physiology or Medicine in 1969.

Luria was born in Turin. He left fascist Italy in 1938, going first to France, where he became a research fellow at the Institut du Radium in Paris, and then to the United States in 1940. From 1943 he taught at a number of universities and in 1959 became a professor at the Massachusetts Institute of Technology (MIT). He founded the MIT Center for Cancer Research, which he directed from 1972 to 1985. For some time he taught a course in world literature to graduate students at MIT and at Harvard Medical School to encourage their involvement in the arts.

Morgan, Ann Haven (1882–1966)

U.S. zoologist who promoted the study of ecology and conservation. She particularly studied the zoology of aquatic insects and the comparative physiology of hibernation. Her *Field Book of Ponds and Streams: An Introduction to the Life of Fresh Water* (1930) attracted amateur naturalists and provided an authoritative taxonomic guide for professionals.

Morgan was born in Waterford, Connecticut, and studied at Wellesley College and Cornell University. From 1912 to 1947, she taught at Mount Holyoke College in Massachusetts, becoming professor in 1918. She spent her summers at a variety of research laboratories, including the Marine Biological Laboratory at Woods Hole in Massachusetts, and also worked at the Tropical Research Laboratory of British Guiana (now Guyana). She was a member of the National Committee on Policies in Conservation Education.

Mullis, Kary B. (1944–)

> *I don't like to do things that are hard. That's the important part of being an inventor—trying to figure out an easy way to do something.*
>
> —*Kary B. Mullis*

Cowinner of the 1993 Nobel Prize for chemistry, Kary Banks Mullis invented a way to synthesize rapidly billions of copies of a particular piece of DNA, the molecule that contains the genetic code. Known as a "polymerase chain reaction" (PCR), the process has had vast implications for such scientific areas as medicine (a new AIDS test), forensics (determining a person's identity from blood samples), paleontology (comparison of extinct and living animals), and biotechnology, since all of these profit from being able to produce huge quantities of a certain length of DNA.

Mullis was born in Lenoir, North Carolina, and spent most of his youth in Columbia, South Carolina. He earned his B.S. in chemistry from Georgia Tech in 1966, after which he moved to Berkeley, California. He finished his Ph.D. in biochemistry at the University of California at Berkeley in 1973.

In 1979, Mullis took a position as a biochemist at the Cetus Corporation in Emeryville, California, in which role he made his discovery of PCR. Dissatisfied with Cetus, he left the company in 1986, the same year he announced his discovery of PCR, and since then has been active as an entrepreneur and freelance consultant.

Nice, Margaret Morse (1883–1974)

U.S. ornithologist who made an extensive study of the life history of the sparrow, work that prompted some fellow scientists to call her the true founder of ethology. One of the few recognized female scientists of her time, she also campaigned against the indiscriminate use of pesticides.

Morse was born in Amherst, Massachusetts, and studied at Mount Holyoke College and Clark University in Worcester, Massachusetts, graduating with a degree in child psychology. She never had an academic appointment.

Her first ornithological research was a detailed study of the birds of Oklahoma. In 1927 she moved to Ohio, where she carried out the study of sparrows that established her as one of the leading ornithologists in the world, recording the behavior of individual birds over a long period of time. A family move to Chicago provided fewer opportunities for Nice to study living birds, so she spent more of her time writing, and became involved in conservation issues.

Nusslein-Volhard, Christine (1942–)

German geneticist who shared the Nobel Prize for Physiology or Medicine in 1995 with U.S. geneticists Edward Lewis and Eric Wieschaus for her work on the genes controlling early embryonic development of the fruit fly *Drosophila melanogaster*. She examined forty thousand random gene mutations for their effect on the fly's development and identified 150 genes. She has since cloned several of those genes and worked out their interactions.

Nusslein-Volhard performed her experiments to identify all of the genes involved in the development of the fruit fly at the European Laboratory for Molecular Biology at Heidelberg. The mutant strains of flies she created have since been worked on by many other developmental biologists.

She was inspired in the late 1970s by the pioneering work of Edward Lewis, who had identified the transformations in the fruit fly that cause substitution of one segment of the body for another. These transformations were found to be the result of mutations in a gene family called the bithorax complex. Genes at the beginning of the complex were found to control anterior body segments, while genes further down the genetic map controlled more posterior body segments. This work was shown to be of particular importance when it was demonstrated that the gene ordering of this complex is conserved in humans.

Since 1986, Nusslein-Volhard has been director of the genetics division of the Max Planck Institute for Developmental Biology, Tübingen. In the 1990s, Nusslein-Volhard embarked on a large-scale project using zebrafish instead of fruit flies to generate and classify genetic variations.

Sanger, Frederick (1918–)

English biochemist who was the first person to win a Nobel Prize for Chemistry twice: the first in 1958 for determining the structure of insulin, and the second in 1980 for work on the chemical structure of genes. Sanger shared the latter prize with two U.S. scientists, Paul Berg and Walter Gilbert, for establishing methods of determining the sequence of nucleotides strung together along strands of RNA and DNA. He also worked out the structures of various enzymes and other proteins.

Sanger was born in Gloucestershire and studied at Cambridge, where he spent his whole career. In 1951 he became head of the Protein Chemistry Division of the Medical Research Council's Molecular Biology Laboratory.

Between 1943 and 1953, Sanger and his coworkers determined the sequence of fifty-one amino acids in the insulin molecule. By 1945 he had discovered a compound, Sanger's reagent (2,4-dinitrofluorobenzene), which attaches itself to amino acids, and this enabled him to break the protein chain into smaller pieces and analyze them using paper chromatography.

From the late 1950s, Sanger worked on genetic material, and in 1977 he and his coworkers announced that they had established the sequence of the more than five thousand nucleotides along a strand of RNA from a bacterial virus called R17. They later worked out the order for mitochondrial DNA, which has approximately seventeen thousand nucleotides.

Sanger, Margaret Louise Higgins (1883–1966)

U.S. health reformer and crusader for birth control. In 1914 she founded the National Birth Control League. She founded and presided over the American Birth Control League from 1921 to 1928; the organization later became the Planned Parenthood Federation of America, and the International Planned Parenthood Federation in 1952.

Sanger was born in Corning, New York. She received nursing degrees from White Plains Hospital and the Manhattan Eye and Ear Clinic. As a nurse, she saw the deaths and deformity caused by self-induced abortions

Undated photograph of Margaret Sanger. (Corbis-Bettmann)

and became committed to providing health and birth-control education to the poor. In 1917 she was briefly sent to prison for opening a public birth-control clinic in Brooklyn in 1916.

Tinbergen, Nikolaas (1907–1988)

Dutch-born British zoologist. He specialized in the study of instinctive behavior in animals. One of the founders of ethology, the scientific study of animal behavior in natural surroundings, he shared a Nobel Prize in 1973 with Austrian zoologists Konrad Lorenz (with whom he worked on several projects) and Karl von Frisch. Tinbergen investigated other aspects of animal behavior, such as learning, and also studied human behavior, particularly aggression, which he believed to be an inherited instinct that developed when humans changed from being predominantly herbivorous to being hunting carnivores.

Tinbergen was born in The Hague and educated at Leiden, where he became professor in 1947. From 1949 he was in England at Oxford, and established a school of animal behavior studies there.

In *The Study of Instinct* (1951), Tinbergen showed that the aggressive behavior of the male three-spined stickleback is stimulated by the red coloration on the underside of other males (which develops during the mating season). He also demonstrated that the courtship dance of the male is stimulated by the sight of the swollen belly of a female that is ready to lay eggs.

In *The Herring Gull's World* (1953), he described the social behavior of gulls, emphasizing the importance of stimulus–response processes in territorial behavior.

Watson, David Meredith Seares (1886–1973)

English embryologist and paleobiologist who provided the first evidence that mammals evolved from reptiles. From the fossilized remains of primitive reptiles and mammals collected on trips to South Africa and Australia between 1911 and 1914, he pieced together the evolutionary line linking reptiles to early mammals.

Watson was born in Higher Broughton, Lancashire. He graduated from Manchester University in chemistry and geology, but began publishing papers on paleobiology while he was still an undergraduate. With the outbreak of World War I, he returned to England to join the Royal Air Force. From 1921 until his retirement in 1965 he was based at University College, London. In 1952 he was Alexander Agassiz visiting professor at Harvard. He was elected to the Royal Society in 1922 and was awarded the society's Darwin medal in 1942 and its Linnaeus medal in 1949.

Watson, James Dewey (1928–)

U.S. biologist. His research on the molecular structure of DNA and the genetic code, in collaboration with English molecular biologist Francis Crick, earned him a shared Nobel Prize in 1962. Based on earlier works, they were able to show that DNA formed a double helix of two spiral strands held together by base pairs.

Watson was born in Chicago and studied at the University of Chicago and at Indiana University. He initially specialized in viruses but shifted to molecular biology and in 1951 he went to the Cavendish Laboratory at Cambridge, England, where he performed the work on DNA with Crick. In 1953, Watson returned to the United States. He became professor at Harvard in 1961 and director of the Cold Spring Harbor Laboratory of Quantitative Biology in 1968 and was head of the U.S. government's Human Genome Project between 1989 and 1992.

James Watson displays his model for the structure of deoxyribonucleic acid (DNA) in 1962. Watson and two British scientists, Harry Compton Crick and Maurice H. F. Wilkins, were the joint recipients of the 1962 Nobel Prize for Physiology or Medicine. (Corbis-Bettmann)

Crick and Watson published their work on the proposed structure of DNA in 1953 to explain how genetic information could be coded. They envisioned DNA replication occurring by a parting of the two strands of the double helix, each organic base thus exposed linking with a nucleotide (from the free nucleotides within a cell) bearing the complementary base. Thus two complete DNA molecules would eventually be formed by this step-by-step linking of nucleotides, with each of the new DNA molecules comprising one strand from the original DNA and one new strand.

Whipple, George Hoyt (1878–1976)

U.S. physiologist whose research interest concerned the formation of hemoglobin in the blood. The son and grandson of physicians, Whipple was born in Ashland, New Hampshire. He earned his B.A. from Yale in 1900 and, having decided to go into medicine, studied for five years at Johns Hopkins University in Baltimore, Maryland. He completed his medical degree there in 1905. Much of Whipple's career was spent at the University of Rochester, where he became professor of pathology and dean of medicine in 1921; he was also long associated with the Rockefeller Foundation.

Whipple's research demonstrated that anemic dogs, kept under restricted diets, responded well to the consumption of liver, and that their hemoglobin quickly regenerated. This work led to a cure for pernicious anemia, which had been fatal up till then. He shared the 1934 Nobel Prize for Physiology or Medicine with U.S. physicians George Minot (1885–1950) and William Murphy (1892–1987).

4

Directory of Organizations

American Institute of Biological Sciences
1444 Eye Street NW, Suite 200
Washington, DC 20005
phone: (202) 628-1500
fax: (202) 628-1509
Web site: http://www.aibs.org

Established in 1947, the institute is a nonprofit scientific organization for the advancement of research, education, professional relations, and public understanding in the biological, medical, environmental, and agricultural sciences. Members' interests span all of biology, basic and applied, from agronomy to zoology. The organization provides scientific services to the public and government, as well as programs and services for its individual members and member societies.

American Ornithologists Union (AOU)
AOU c/o Division of Birds MRC 116
National Museum of Natural History
Washington, DC 20560
Web site: http://pica.wru.umt.edu/AOU/AOU.html

Organization for the scientific study of birds. The AOU, established in 1883, is primarily a professional organization; however, its membership includes many amateurs dedicated to the advancement of ornithological science.

American Physiological Society
9650 Rockville Pike
Bethesda, MD 20814-3991
phone: (301) 530-7118
fax: (301) 571-8305
Web site: http://www.faseb.org/aps/

Nonprofit scientific society established in 1887 devoted to education, scientific research, and the dissemination of information in the physiological sciences. Most members have doctoral degrees in physiology or medicine (or other health professions) or both.

American Society for Biochemistry and Molecular Biology

9650 Rockville Pike
Bethesda, MD 20814-3996
phone: (301) 530-7145
fax: (301) 571-1824
email: asbmb@asbmb.faseb.org
Web site: http://www.faseb.org/asbmb/

Nonprofit scientific and educational organization for biochemistry and molecular biology. Most members teach and conduct research at colleges and universities. Others conduct research in various government laboratories, research institutions, and industry.

American Society for Microbiology

1325 Massachusetts Avenue
Washington, DC 20005
phone: (202) 737-3600
Web site: http://www.asmusa.org

Society established in 1899 representing twenty-four disciplines of microbiological specialization with a division for microbiology educators.

American Society of Parasitologists

George D. Cain, ASP Secretary-Treasurer
Department of Biological Sciences
University of Iowa
Iowa City, IA 52242 (608) 263-1107
fax: (608) 263-6573
Web site: http://www-museum.unl.edu/asp/

Society of scientists established in 1924 from industry, government, and academia who are interested in the study and teaching of parasitology, the study of parasites. Members have contributed to the development of parasitology as a discipline, and to primary research in systematics, medicine, molecular biology, immunology, physiology, ecology, and biochemistry. The society publishes the *Journal of Parasitology*.

Anatomical Society of Great Britain and Ireland
Department of Human Anatomy
University of Oxford, South Parks Road
Oxford OX1 3QX, United Kingdom
phone: +44 (0)1865 272165
fax: +44 (0)1865 272420
email: @human-anatomy.oxford.ac.uk
Web site: http://www.sm.ic.ac.uk/

Scientific organization founded in 1887 for the promotion and development of the anatomical and related sciences, serving principally anatomists in the United Kingdom and the Republic of Ireland, but also has members worldwide. The *Journal of Anatomy*, covering all aspects of normal human and comparative anatomy, is published on behalf of the society by Cambridge University Press.

Australian Institute of Marine Science (AIMS)
PMB No.3, Townsville MC
Queensland, 4810 Australia
phone: +61 (07) 4753 4444
fax: +61 (07) 4772 5852
Web site: http://www.aims.gov.au/

Scientific and technological research organization for the sustainable use and protection of the marine environment. AIMS was established by the Commonwealth government in 1972.

Australian Society for Microbiology (ASM)
Unit 23, 20 Commercial Road
Melbourne, Victoria 3004, Australia
phone: +61 3 9867 8699
fax: +61 3 9867 8722
email: TheASM@asm.auz.com
Web site: http://home.vicnet.net.au/~asm/welcom2.htm

Professional, nonprofit organization, founded in 1959 to further the science of microbiology in Australia. ASM has branches in each state and the Australian Capital Territory, and each branch organizes lectures, workshops, and other scientific activities. ASM publications include its official journal *Microbiology Australia*, published five times a year, and the annual volume *Recent Advances in Microbiology*. A federation of Asia Pacific Microbiology Societies has been formed under the auspices of ASM for the further dissemination of relevant microbiology issues, especially the identification, treatment, eradication, and management of infectious diseases.

Biodiversity Forum
8000 Towers Crescent Drive, Suite 1350
Vienna, VA 22182
phone: (703) 847-3686
fax: (703) 760-7899
Web site: http://www.worldcorp.com/biodiversity

Nonprofit organization dedicated to the international conservation of wildlife, habitat, and biological diversity, specializing in international conservation science and policy. The organization promotes public understanding of issues relating to the conservation of worldwide biological diversity; encourages cooperation among governments, international organizations, and the private sector in developing methods for the sustainable use of biological resources; and collects and disseminates information relating to international treaties and conventions designed to regulate the use of biological resources. It also monitors developments concerning the conservation of biological resources governed at the international level, principally the developments in the Convention on International Trade in Endangered Species of Wild Fauna and Flora (CITES) and the Convention on Biological Diversity (CBD).

British Entomological Society and Natural History Society (BENHS)
The Pelham-Clinton Building
Dinton Pastures Country Park, Davis Street
Hurst, Reading RG10 0TH, Berkshire, United Kingdom
Web site: http://ourworld.compuserve.com/homepages/pyo/intro2. htm

Organization for the promotion and advancement of research in entomology, the study of insects. Founded in 1872 as the South London Entomological and Natural History Society, BENHS places emphasis on the conservation of fauna and flora in the United Kingdom, and the protection of wildlife throughout the world. The society publishes key reference works on British entomology, and its quarterly *British Journal of Entomology and Natural History* includes scientific papers, articles, and observations.

CAB International
Wallingford
Oxon OX10 8DE, United Kingdom

phone: +44 (0)1491 832111
fax: +44 (0)1491 833508
email: cabi@cabi.org
Web site: http://www.cabi.org/

International nonprofit organization for scientific research and development. It was established in 1913 as the Imperial Agricultural Bureaux (IAB) and renamed the Commonwealth Agricultural Bureaux in 1948; it became fully international in 1985, with membership open to any country, and was reconstituted as CAB International in line with its new status. It has three principal areas of activity: Bioscience, undertaking research and training in biological pest management, biodiversity, biosystematics, and the environment; Information for Development, assisting developing countries to acquire and manage scientific information; and Publishing, producing publications in all fields of agriculture and related fields.

Carnegie Institution of Washington
P Street NW
Washington, DC 20005
phone: (202) 387-6400
Web site: http://www.ciw.edu/intro.htm www.ciw.edu/Geo_seismo.html

Nonprofit organization founded and funded by Andrew Carnegie and incorporated by an Act of Congress in 1904. Carnegie, who had more money than the U.S. treasury at the time, endowed the institution with $10 million, envisioning a research facility that would "encourage, in the broadest and most liberal manner, investigation, research, and discovery, and the application of knowledge to the improvement of mankind." Research at the institute includes the earth sciences, astronomy, and biology, with an emphasis on predoctoral and postdoctoral education. As well as their introductory Web site, the institute also maintains a Web listing of links to seismology, volcano, and earthquake sites, as well as links to major institutions and universities.

Center for Plant Conservation
P.O. Box 299
Missouri Botanical Garden
St. Louis, Missouri 63166-0299
phone: (314) 577-9450
fax: (314) 577-9465
email: cpc@mobot.org
Web site: http://www.mbot.org/CPC/

Organization that aims to prevent the extinction of plants native to the United States. It coordinates the development and maintenance of the National Collection of Endangered Plants, a living collection of rare native U.S. plant taxa maintained in protective cultivation and its associated conservation, research, and education programs. The organization maintains a comprehensive database on the biology, horticulture, and conservation status of rare plants of the United States, through which it can track the status of taxa of concern, and encourages ethical scientific research on the rare plants of the United States.

European Molecular Biology Laboratory (EMBL)
Postfach 10.2209
69012 Heidelberg
Federal Repuplic of Germany
phone: +49 6221 3870 (sic)
fax: +49 6221 387306
email: @EMBL-Heidelberg.de
Web site: http://www.embl-heidelberg.de/

Scientific institution for research into molecular and cell biology. It was established in 1974 and consists of a main laboratory in Heidelberg, Germany, and three outstations: Hamburg, Germany; Grenoble, France; and Hinxton, near Cambridge, United Kingdom. The Hinxton outstation specializes in bioinformatics research, while Hamburg and Grenoble have highly intensive X-ray and neutron radiation equipment for structural studies. EMBL is also a teaching and training center for molecular biology, and is supported by fourteen European countries and Israel.

European Molecular Biology Organization (EMBO)
Postfach 1022.40
D-69012 Heidelberg
Federal Republic of Germany
phone: +49 6221 383031
fax: +49 6221 384879
email: EMBO@EMBL-Heidelberg.de
Web site: http://www.embo.org/

International body for the promotion of molecular biology studies in Europe. It is funded by contributions from its twenty-three member states, which together form the European Molecular Biology Conference (EMBC). EMBO produces the scientific journal of molecular biology *EMBO Journal*, published twice monthly by the Oxford University Press.

Federation of European Microbiological Societies
FEMS Central Office
Poortlandplein 6, 2628 BM Delft
The Netherlands
phone: +31 (15)278 5604
fax: +31 (15)278 5696
email: fems@tudelft.nl
Web site: http://www.elsevier.nl:80/inca/homepage/sah/fems/menu.htm

International scientific organization for the promotion of microbiology in Europe. Linking some thirty-eight microbiological societies throughout Europe, the federation encourages joint activities that facilitate communication among microbiologists. It supports meetings and laboratory courses, provides fellowships and grants for young scientists, and publishes journals and books in related fields.

Field Museum of Natural History
Roosevelt Road at Lake Shore Drive
Chicago, IL 60605
phone: (312) 922-9410
Web site: http://www.fmnh.org

Educational institution established in 1893 concerned with the diversity and relationships in nature and among cultures. It provides collection-based research and learning for greater public understanding and appreciation of the world in which we live.

Food and Agriculture Organization of the United Nations
Viale delle Terme di Caracalla
00100 Rome, Italy
phone: 39(6) 57051
fax: 39(6) 57053152
telex: 625852/625853/610181 FAO I
telegrams: FOODAGRI ROME
Web site: http://www.fao.org

United Nations organization with a mandate to raise levels of nutrition and standards of living, to improve agricultural productivity, and to better the condition of rural populations. The organization, which comprises 174 member nations plus the European Union as a member organization, offers direct development assistance; collects, analyzes, and disseminates information; provides policy and planning advice to governments; and acts as an international forum for debate on food and agriculture issues. It is active in land and water development, plant and animal production, forestry, fisheries, economic and social policy, investment, nutrition, food standards, commodities, and trade. It also plays a major role in dealing with food and agricultural emergencies.

Human Genome Mapping Project Resource Center (HGMP-RC)
UK MRC HGMP Resource Center
Hinxton, Cambridge CB10 1SB, United Kingdom
phone: +44 (0)1223 494500
fax: +44 (0)1223 49451
email: admin@hgmp.mrc.ac.uk
Web site: http://www.hgmp.mrc.ac.uk/

British scientific body for the provision of biological and data resources and services to the medical research community working on the Human Genome Program. Funded by the Medical Reseach Council (MRC), the HGMP-RC is located at the Hinxton Genome Campus, with the Sanger Center and the European Bioinformatics Institute. It provides access to genomic libraries and updated databases, and has its own, newly formed research division.

Human Genome Program, U.S. Department of Energy (DOE)
Human Genome Management Information System
Oak Ridge National Laboratory
1060 Commerce Park MS 6480
Oak Ridge, TN 37830
fax: (423) 574-9888
email: mansfieldbk@ornl.gov
Web site: http://www.ornl.gov/hgmis

Program to work on the U.S. Human Genome Project in cooperation with the National Human Genome Research Institute of the National Institutes of Health. The U.S. project is part of a larger international endeavor to characterize the genomes of humans and several model organisms. Other genome programs include the DOE microbial genome program, a project to characterize microbes of environmental or industrial interest. The DOE Human Genome

Program includes research projects at universities, DOE genome centers, DOE-owned national laboratories, and other research organizations.

The Institute for Genomic Research (TIGR)
9712 Medical Center Drive
Rockville, MD 20850
phone: (301) 838-0200
fax: (301) 838-0208
Web site: http://www.tigr.org

Research institute with interests in structural, functional, and comparative analysis of genomes and gene products in viruses, eubacteria, pathogenic bacteria, archaea, and eukaryotes (both plant and animal), including humans.

Institute of Biology
20–22 Queensberry Place
London SW7 2ZY, United Kingdom
phone: +44 (0)171 581 8333
fax: +44 (0)171 823 9409
email: info@iob.primex.co.uk
Web site: http://www.iob.org/

Chartered professional body for U.K. biologists; founded in 1950 to promote biology as a science and to develop applications of biology. It has more than seventy affiliated societies and 16,500 members.

Institute of Biomedical Science
12 Coldbath Square
London EC1R 5HL, United Kingdom
phone: +44 (0)171 713 0214
fax: +44 (0)171 436 4946
email: mail@ibms.org
Web site: http://www.ibms.org/

British professional body for biomedical scientists. Its aim is to promote and develop biomedical science and its practitioners. It was founded in 1912 and changed its name from the Institute of Medical Laboratory Sciences in 1994; it has branches in Hong Kong, Cyprus, and Gibraltar.

Institute of Cancer Research (ICR)
17a Onslow Gardens
London SW7 3AL, United Kingdom
phone: +44 (0)171 352 8188
fax: +44 (0)171 349 9765
email: m&a@icr.ac.uk
Web site: http://www.icr.ac.uk/

British scientific public sector organization for research into the eradication of cancer. In partnership with the Royal Marsden NHS Trust, it is the largest cancer center in Europe and is an associated institution of the University of London. The majority of the ICR's six hundred staff and students are directly engaged in research. Its largest single sponsor is the Cancer Research Campaign, and considerable funding is received from the Medical Research Council, the Royal Marsden Hospital, and the Leukemia Research Fund. The institute conducts the fight against cancer through research and development; training health professionals and scientific staff; and providing high quality patient care.

International Center for Genetic Engineering and Biotechnology (ICGEB)
AREA Science Park
Padriciano 99, 34012
Trieste, Italy
phone: +39 40 37571
fax: +39 40 226555
email: icgeb@icgeb.trieste.it
Web site: http://www.icgeb.trieste.it/

International organization, composed of forty-one member states, established to promote the safe use of biotechnology worldwide, with special regard to the needs of the developing world. The two main ICGEB laboratories are located in Trieste, Italy, and in New Delhi, India. The organization hosts a network of national laboratories in member states. Specific research programs include the study of human genetic diseases, the genetic manipulation of plants, and the production of novel malaria and hepatitis vaccines.

International Federation for Medical and Biological Engineering
IFMBE Secretariat
Department of Biomedical Engineering
Huddinge University Hospital, SE-141 86

Huddinge, Sweden
phone: +46 8 585 80852
fax: +46 8 585 86290
Web site: http://www.iupesm.org/ifmbe.html

> International scientific organization for the research and development
> of medical and biological engineering. It was founded in 1959 as the
> International Federation for Medical Electronics and Biological Engi-
> neering, and now has forty-three affiliated organizations.

International Union for the Conservation of Nature
IUCN World Headquarters
Rue Mauverney 28
CH 1196 Gland
Switzerland
phone: +41 22 9990001
fax: +41 22 9990002
email: mail@hq.iucn.org
Web site: http://www.iucn.org/

> International organization established in 1948 for the conservation of
> the integrity and diversity of nature. It comprises a global network of
> institutions and organizations, including governments and government
> agencies, and gives advice and assistance to governments, organiza-
> tions, and local communities in devising and implementing conserva-
> tion strategies.

**International Union of Biochemistry and Molecular
Biology (IUBMB)**
Technical University Berlin
Max-Volmer-Institute for Biophysical Chemistry and Biochemistry
Franklinstrasse 29, D-1058
Berlin, Germany
phone: (49 30) 3142 4205
fax; (49 30) 3142 4783
email: kleinkauf@chem-tu-berlin.de
Web site: http://www.lmcp.jussieu.fr/icsu/Membership/SUM/iubmb.html

> Organization formed in 1955 to promote international cooperation
> in biochemistry. Its aim is to promote high standards of biochemistry
> through research, discussion, publication, and standardization. The
> IUBMB sponsors the International Congress of Biochemistry and

Molecular Biology every three years, and in the intervening years IUBMB conferences take place.

International Union of Biological Sciences
51 Boulevard de Montmorency
75016 Paris, France
phone: +33 1 4525 0009
fax: +33 1 4525 2029
email: iubs@paris7.jussieu.fr
Web site: http://www.iubs.org/

Nonprofit scientific organization for study and research in biological sciences. It was established in 1919 and is a founding member of the International Council of Scientific Unions (ICSU). The union has close relationships with the United Nations Educational, Scientific and Cultural Organization (UNESCO), the European Commission, and other organizations in the development of joint collaborative research and training programs related to biological sciences.

IUCN/World Conservation Union
Address: Rue Mauverney 28
CH-1196 Gland, Switzerland
phone: (41 22) 999 0001
fax: (41 22) 999 0002
email: oedinfo@unep.org
Web site: http://w3.iprolink.ch/iucnlib/

International conservation organization established in 1948. It links governments and governmental and nongovernmental agencies with scientists and experts to encourage a worldwide approach to conservation. The IUCN (International Union for the Conservation of Nature) works with six global commissions formed to assist in the protection of nature in the following areas: species survival, protected areas, education and communication, environmental law, ecosystem management, and environmental economics and social policy. IUCN publications include the *Red List of Threatened Animals* and the *Red Data Books*.

Linnean Society of London
Burlington House
Piccadilly, London
W1V 0LQ, United Kingdom
Web site: http://www.linnean.org.uk/

British scientific body for the study of diversity in evolution, ecology, and systematics. Founded in 1788, it is the oldest extant scientific society in the world devoted to natural history and is named for the Swedish naturalist Carl von Linné (1707–1778). The society promotes study of all aspects of biology, especially the diversity and interrelationships of organisms, which involves the examination and collation of scientific evidence from the fields of genetics, ecology, anatomy, physiology, biochemistry, and paleontology. The society's *Biological Journal of the Linnean Society* is the oldest of its kind in the world. First produced in 1791, the original papers on natural selection by Darwin and Wallace appeared in this journal in 1858. It devotes extensive coverage to the processes of organic evolution and also publishes papers on theoretical, genetic, and population biology. The *Botanical Journal of the Linnean Society* publishes research papers on the plant sciences, and the *Zoological Journal of the Linnean Society* publishes original papers on zoology, and related research areas including systematics, comparative anatomy, functional zoology, ecology, behavior, and zoogeography.

Medical Research Council
20 Park Crescent
London W1N 4AL, United Kingdom
phone: +44 (0)171 636 5422
fax: +44 (0)171 436 6179
email: tony.helm@headoffice.mrc.ac.uk
Web site: http://www.mrc.ac.uk/

British research organization for the promotion of strategic and applied research in the biomedical sciences. Results of the council's research projects are disseminated with the aim of maintaining and improving human health in the United Kingdom.

National Center for Biotechnology Education (NCBE)
School of Animal and Microbiological Sciences
The University of Reading, Whiteknights
P.O. Box 228
Reading RG6 6AJ, United Kingdom
phone: +44 (0)118 9873743
fax: +44 (0)118 9750140
Web site: http://www.ncbe.reading.ac.uk/

Independent British organization for the provision of information, training, and resources for schools and colleges, industry, professional organizations, and the general public. NCBE was founded in 1985 to support the teaching of biotechnology in schools and colleges and was the first of its type in Europe; it is part of the Department of Microbiology at the University of Reading. Initially funded via a U.K. government grant, the center is now self-funded by charging for its services.

National Center for Biotechnology Information
National Library of Medicine
Building 38A, Room 8N805
Bethesda, MD 20894
phone: (301) 496-2475
fax: (301) 480-9241
email: info@ncbi.nlm.nih.gov
Web site: http://www.ncbi.nlm.nih.gov

Organization developing information technologies to aid in the understanding of the molecular and genetic processes that control health and disease. It creates automated systems for storing and analyzing knowledge about molecular biology, biochemistry, and genetics; carries out research into computer-based information processing for analyzing the structure and function of biologically important molecules; facilitates the use of databases and software by biotechnology researchers and medical personnel; and coordinates the collection of biotechnology information worldwide.

National Center for Genome Resources
1800-A Old Pecos Trail
Santa Fe, NM 87505
phone: (505) 982-7840
fax: (505) 995-4432
Web site: http://www.ncgr.org

Nonprofit organization providing central collection and analysis of genetic information from around the world, including the Human Genome Project. The center's library of data, connections, and information is available through the Genome Sequence DataBase, a bioinformatics resource for researchers, and the Genetics and Public Issues program, which educates public and professional audiences about genetic issues.

National Institute for Medical Research
The Ridgeway, Mill Hill
London NW7 1AA, United Kingdom

phone: +44 (0)181 959 3666
fax: +44 (0)181 906 4477
email: @nimr.mrc.ac.uk;
Web site: http://www.nimr.mrc.ac.uk/

British research organization established in 1920 under the auspices of the Medical Research Council (MRC). It covers a broad range of basic medical research, currently organized in four major groupings; genes and cellular controls; infection and immunity; neurosciences; and structural biology. Funded mainly by the MRC, it is also supported by medical research charities and industrial and commercial companies.

National Museum of Natural History, Smithsonian Institution
National Mall
Washington, DC
phone: (202) 357-2700
Web site: http://www.usaemb.pl/digilib/smith/0/NA113.htm

Museum of earth history, human cultures, and the diversity of the natural world. The museum collects, preserves, studies, and displays specimens from the natural world and objects made by its inhabitants. It carries out research and maintains the national collections, which comprise more than 120 million scientific specimens and cultural artifacts from around the world. It runs programs and projects to increase public understanding of the natural world.

National Park Service of the United States
The Department of the Interior
National Park Service
Office of Public Inquiries
P.O. Box 37127, Room 1013
Washington, DC 20013-7127
phone: (202) 208 4747
email: nps_webmaster@nps.gov
Web site: http://www.nps.gov

Service promoting and regulating the use of the U.S. national parks to conserve the natural lands, wildlife, and historic structures for the enjoyment of future generations. The National Park System of the United States comprises 376 areas covering more than eighty-three million acres in forty-nine states, the District of Columbia, American Samoa, Guam, Puerto Rico, Saipan, and the Virgin Islands. There are three principal categories used in classification: natural areas, historical areas, and recreational areas, each of which have special recognition and protection in accordance with acts of Congress.

Pathological Society of Great Britain and Ireland
2 Carlton House Terrace
London SW1Y 5AF, U
phone: +44 (0)171 976 1260
fax: +44 (0)171 976 1267
email: administrator@pathsoc.org.uk
Web site: http://www.pathsoc.org.uk/

Scientific organization for the advancement of pathology and allied sciences related to diseases and disease processes. The society publishes two international monthly journals: the *Journal of Pathology* and the *Journal of Medical Microbiology*. It also sponsors students (both undergraduate and postgraduate), and has fellowships for the medical and allied professions.

Research School of Biological Sciences (RSBS)
The Australian National University
GPO Box 475
Canberra City, ACT 2601, Australia
phone: +61 (0)2 6249 2999
fax: +61 (0)2 6249 4891
email: @rsbs.anu.edu.au
Web site: http://biology.anu.edu.au/

One of eight research schools constituting the Institute of Advanced Studies at the Australian National University. It is a leading center of biological research and graduate training in selected areas of biology: molecules, cells, organisms, ecosystems, and the biosphere. The RSBS publishes scientific papers and the biannual journal *Biologic*, aimed at secondary school science and general interest readers.

Roslin Institute
Roslin, Midlothian
EH25 9PS, United Kingdom
phone: +44 (0)131 527 4200
fax: +44 (0)131 440 0434
email: roslin.library@bbsrc.ac.uk
Web site: http://www2.ri.bbsrc.ac.uk/

International center for research on molecular and quantitative genetics of farm animals and poultry science. Its major research programs include the biology of reproduction, developmental biology and growth, and animal welfare and behavior. It is an independent nonprofit organization, sponsored by the Biotechnology and Biological Sciences Research Council (BBSRC).

Royal Society
6 Carlton House Terrace
London SW1Y 5AG, United Kingdom
phone: +44 (0)171 839 5561
fax: +44 (0)171 930 2170
email: ezmb013@mailbox.ulcc.ac.uk
Web site: http://www.royalsoc.ac.uk/

British independent academy, founded in 1660, for the promotion of natural and applied sciences. It provides a broad range of services for the scientific community in the national interest and supports international scientific exchange. The society's publications include journals, which provide research findings and authoritative reviews; *Notes and Records,* the Royal Society journal on the history of science; publications of record; reports on scientific studies; submissions to government; and reports on U.K. science.

Sanger Center
Wellcome Trust Genome Campus
Hinxton, Cambridge CB10 1SA, United Kingdom
phone: +44 (0)1223 834244
fax: +44 (0)1223 494919
email @sanger.ac.uk
Web site: http://www.sanger.ac.uk

British research establishment for the sequencing, mapping, and interpretation of the human genome. It was established in 1992 by the Wellcome Trust and the Medical Research Council (MRC), and is situated at the Hinxton Genome Campus. Renewed funding from the Wellcome Trust enables an increased production rate, and it is estimated that the completion of the entire human genome sequence by an international collaboration, including the Sanger Center, will be accomplished by 2005.

Scientific Committee on Problems of the Environment (SCOPE)
51 Boulevard de Montmorency
75016 Paris, France
phone: (33 1) 4525 0498
fax: (33 1) 4288 146
email: scope@paris7.jussieu.fr
Web site: http://www.lmcp.jussieu.fr/icsu/Structure/scope.html

International organization formed in 1969 to provide information to organizations and agencies involved in the study of the environment. To
this end, SCOPE assembles data and reviews and assesses changes to
the environment and also evaluates the methods of measurement used.

Society for Developmental Biology
9650 Rockville Pike
Bethesda, MD 20814-3998
phone: (301) 571-0647
fax: (301) 571-5704
Web site: http://sdb.bio.purdue.edu/

Nonprofit society dedicated to the advancement of scientific and educational excellence in developmental biology, and to public understanding of developmental biology. It is the professional society associated
with the journal *Developmental Biology*, published by Academic Press.

Society for Endocrinology
17/18 The Courtyard
Woodlands, Bradley Stoke
Bristol B532 4NQ, United Kingdom
phone: +44 (0)1454 619347
fax: +44 (0)1454 616071
email: info@endocrinology.org
Web site: http://www.endocrinology.org/

Scientific organization established in 1946 to promote advances in endocrinology. A founding member of the British Endocrine Societies, it
is the major non-U.S. society of its kind. The society's publications
include the monthly *Journal of Endocrinology*, and the bimonthly *Journal of Molecular Endocrinology*.

Society for Experimental Biology
Burlington House, Piccadilly
London W1V 0LQ, United Kingdom

phone: +44 (0)171 439 8732
fax: +44 (0)171 287 4786
email: v.wragg@sebiol.demon.co.uk
Web site: http://www.demon.co.uk/SEB/

British nonprofit scientific organization for the promotion of all aspects of experimental biology. Its three main sections include animal, cell, and plant biology. The society holds an annual symposium on a specialized topic, the proceedings of which are published; it also sponsors the monthly *Journal of Experimental Biology,* which publishes research papers in the plant sciences.

Society for General Microbiology (SGM)
Marlborough House
Basingstoke Road
Spencer's Wood
Reading RG7 1AE, United Kingdom
phone: +44 (0)118 988 1800
fax: +44 (0)118 988 5656
email: web.admin@socgenmicrobiol.org.uk
Web site: http://www.socgenmicrobiol.org.uk/

British scientific organization for the advancement of the science of microbiology. It was founded in 1945, holding its first scientific meeting in Cambridge the same year; Sir Alexander Fleming was elected as its first president. The SGM supports microbiology worldwide through its various grant schemes, prize lectures, and awards. Its publications include the *Journal of General Microbiology,* first published in 1947, and the *Journal of General Virology,* founded in 1966.

Society for the Study of Reproduction
1526 Jefferson Street
Madison, WI 53711-2106
phone: (608) 256-2777
fax: (608) 256-4610
email: ssr@ssr.org
Web site: http://www.ssr.org/~ssr/ssr/

Society established in 1967 to promote the study of reproduction by fostering interdisciplinary communication among scientists, holding conferences, and publishing research. Members are basic scientists, medical and veterinary physicians, trainees in graduate and professional schools, and others engaged in research, education, and training in fields relevant to reproductive biology. The research focuses on problems in human and animal reproduction as it relates to medicine, agriculture, and basic biology.

United Nations Educational, Scientific and Cultural Organization (UNESCO)
7 Place de Fontenoy
75352 Paris 07 SP
Paris, France
phone: (33 1) 4568 1000
fax: (33 1) 4567 1690
Web site: http://www.unesco.org/

International organization established in 1945. Its aim is to contribute to world security and peace through education, science, culture, and communication among nations. UNESCO performs five main functions toward this aim: prospective studies, the sharing of knowledge, standard setting, expertise, and exchange of specialized information. It currently has 186 member states.

U.S. Fish and Wildlife Service, Division of Endangered Species
Mail Stop 452ARLSQ
1849 C Street NW
Washington, DC 20240
Web site: http://www.fws.gov/r9endspp/endspp.html

Agency of the U.S. Department of the Interior that works to conserve, protect, and enhance fish and wildlife and their habitats for the continuing benefit of the American people. It lists, reclassifies, and de-lists species under the Endangered Species Act; provides biological opinions to federal agencies on their activities that may affect listed species; oversees recovery activities for listed species; and provides for the protection of important habitat.

Wellcome Trust
The Wellcome Building
183 Euston Road
London NW1 2BE, United Kingdom
phone: +44 (0)171 611 8888
fax: +44 (0)171 611 8545
email: infoserve@wellcome.ac.uk
Web site: http://www.wellcome.ac.uk/

British-based medical research charity. Established in 1936 by the will of Sir Henry Wellcome, founder of the Burroughs-Wellcome pharmaceutical company, it is the largest nongovernmental source of funds for biomedical research in Europe. The trust predominantly supports research in the United Kingdom, but also devotes substantial funds to overseas research; its International Program runs a number of overseas funding schemes. Henry Wellcome took a personal interest in health problems in the tropics, and this is reflected in the activities of the Tropical Medicine program; the Populations Studies program was set up in 1994 in recognition of the major health problems associated with the growth of human populations.

World Conservation Monitoring Center
219 Huntingdon Road
Cambridge, CB3 0DL, United Kingdom
phone: +(44 1223) 277314
fax: +(44 1223) 277136
email: info@wcmc.org.uk
Web site: http://www.wcmc.org.uk

Organization providing information services on the conservation and sustainable use of the world's living resources and supporting the development of information systems. It runs a database of information on threatened species and others of conservation concern. Part of this database is used to generate the IUCN Red List of Threatened Animals, and this information is available on the Web site in interactive format.

World Wide Fund for Nature
1250 24th Street NW
Washington, DC 20037
phone: (800) 225-5993
Web site: http://www.worldwildlife.org/

WWF, formerly the World Wildlife Fund, international organization established in 1961 to raise funds for conservation by public appeal. Projects include conservation of particular species, for example, the tiger and giant panda, and special areas, such as the Simen Mountains, Ethiopia.

5

Selected Works
for Further Reading

Attenborough, David. *Life on Earth*. Boston: Little, Brown, 1979. 319 pages. ISBN: 0316057452.

One of the best-written introductions to the variety of life on earth—and beautifully illustrated too. Based on his award-winning television series of the same name, *Life on Earth* takes readers through the whole "story of evolution," beginning with the primeval soup and ending up with us.

Austin, C. R., and R. V. Short, eds. *Reproduction in Mammals*. New York: Cambridge University Press, 1982–1986. 5 vols. ISBN: 0521246288.

A scholarly, comprehensive, and authoritative reference series justly famous for its clear and interesting presentation of up-to-date data and theory. Early volumes cover germ cells and fertilization, hormones in reproduction, embryonic and fetal development, reproductive patterns, and the manipulation of reproduction.

Baker, Robin. *Sperm Wars*. New York: BasicBooks, 1996. 319 pages. ISBN: 0465081797.

Since the 1970s, biologists have been fascinated by the biological and evolutionary implications of sperm from different males competing for fertilization of the egg in the female reproductive tract. This is the first popular book on the phenomenon in humans. It summarizes an immense amount of information, all carefully documented. The treatment of the topic is iconoclastic and provocative.

Bonner, John Tyler. *Life Cycles*. Princeton, NJ: Princeton University Press, 1993. 209 pages. ISBN: 0691033196.

The author is an evolutionary biologist who has devoted his life to the study of slime molds. Writing with clarity and humor, he sets reproduction in the context of the life cycle—a linkage of evolution, development, and the complex activities of adult organisms. Filled with wonderful insights and interesting examples.

Calvino, Italo. *Time and the Hunter.* London: Picador, in association with Jonathan Cape, 1993, 1969. 152 pages. ISBN: 0330319094.

Short stories inspired by science. Several of the stories spring from the wonders of cell division, and have their own kind of truth. Translated from the Italian by William Weaver.

Carson, Rachel. *Silent Spring.* Boston: Houghton Mifflin; Cambridge, MA: Riverside Press, 1962. 368 pages.

The one book that likely more than any other kick-started today's environmental movement. Rachel Carson, a marine biologist and environmentalist, drafted this passionate exposé about the impact of pesticides on the environment when she was already seriously ill with cancer. The book caused a furor, and Carson spent her last months under bitter attack by industry representatives. She helped change the outlook of a generation. Beautifully written and cogently argued, the work remains fresh almost four decades later.

Clark, David, and Lonnie Russell. *Molecular Biology Made Simple.* Vienna, IL: Cache River Press, 1997. 470 pages. ISBN: 0962742295.

Aimed at students, but with a lively writing style that gives it appeal to a wider audience.

Cohen, Jack. *Reproduction.* London; Boston: Butterworth, 1977. 356 pages. ISBN: 0408707984.

A standard, well-illustrated introduction that gathers together information on all aspects of biological reproduction.

Cronin, Helena. *The Ant and the Peacock.* Cambridge; New York: Press Syndicate of the University of Cambridge, 1991. 490 pages. ISBN: 052132937X.

A beautifully written account of two important evolutionary topics, altruism and sexual selection, tracing them from their origins in the writings of Darwin and Wallace to the sophisticated theories of mod-

ern times. Interesting for its perspectives linking ecology with historical movements.

Davies, Nicholas. *Dunnock Behavior and Social Evolution.* Oxford
[England]; New York: Oxford University Press, 1992. 272 pages. ISBN:
0198546742.

Fifty years after Lack's *Life of the Robin,* this is a brilliantly written account of the insights that arise from combining detailed fieldwork on
a species—*Prunella modularis,* a hedge sparrow—with modern evolutionary theory.

Dawkins, Richard. *The Blind Watchmaker.* New York: Norton, 1986. 332
pages. ISBN: 0393022161.

Argues that Darwinian natural selection is the only known theory that
could, in principle, explain adaptive complexity.

———. *The Selfish Gene.* New York: Oxford University Press, 1976. 224
pages. ISBN: 019857519X.

The author's first book and still his most famous. An engrossing look
at evolution and behavior from a gene's-eye perspective, expounded
with clarity, wit, and verve. Forceful and persuasive.

Dyson, George. *Darwin amongst the Machines.* Reading, MA: Perseus
Books, 1998. 286 pages. ISBN: 0738200301.

Concerns both biological and artificial life.

Edey, Maitland A., and Donald C. Johanson. *Blueprints.* Boston: Little,
Brown, 1989. 418 pages. ISBN: 0316210765.

A brilliantly written and exciting account that shows just how much
has been achieved by geneticists.

Evans, Howard Ensign. *Life on a Little-Known Planet.* New York: Lyons
& Burford, 1993. 330 pages. ISBN: 1558212493.

Readable introduction to insects, their classification, biological diversity, and effects on people. The author's love of his subject shines
through.

Feduccia, Alan. *The Age of Birds.* Cambridge, MA: Harvard University
Press, 1980. 196 pages. ISBN: 0674009754.

The evolutionary history of birds is a topic of great current research, but this remains the best introduction to the subject.

Foelix, Rainer F. *Biology of Spiders.* New York: Oxford University Press; [Stuttgart] Georg Thieme Verlag, 1996. 330 pages. ISBN: 0195095936.

Entertaining reading for anyone with a basic grasp of biological terminology, but rather like a textbook in its layout.

Ford, Norman D. *When Did I Begin?* Cambridge [Cambridgeshire]; New York: Cambridge University Press, 1988. 217 pages. ISBN: 052134428X.

In-depth analysis of an important ethical problem: when, exactly, does a human embryo become an individual?

Fowler, Cary, and Pat Mooney. *Shattering: Food, Politics and the Loss of Genetic Diversity.* Tucson: University of Arizona Press, 1990. 278 pages. ISBN: 0816511543.

Biodiversity loss doesn't only affect natural systems; during the current century we have lost hundreds of traditional crop strains and livestock breeds. This meticulously researched and readable book looks at what has happened, how genetic diversity has been lost in food production, and why it matters.

Gadagkar, Raghavendra. *Survival Strategies.* Cambridge, MA: Harvard University Press, 1997. 196 pages. ISBN: 0674170555.

Evolutionary explanations for survival strategies elegantly written and beautifully illustrated.

Goodall, D. W., ed. *Ecosystems of the World.* Amsterdam, New York: Elsevier Scientific Pub. Co.

Monumental, multivolume compendium series with extensive bibliographies.

Goodwin, Brian. *How the Leopard Changed Its Spots.* New York: C. Scribner's Sons, 1994. 252 pages. ISBN: 0025447106.

As an outspoken critic of those biological models that see the gene as all-powerful, Goodwin here presents his own ideas about the nature of development.

Gould, Stephen Jay. *The Mismeasure of Man.* New York: Norton, 1981. 352 pages. ISBN: 0393014894.

The Harvard scientist and science writer turns his skills to examining the history of attempts to link intelligence to genetics.

———. *The Panda's Thumb.* New York: Norton, 1980. 343 pages. ISBN: 0393013804.

A collection of essays by a master science writer able to breathe life into facts both general and particular. Here, Gould reflects on the evolution of the panda, investigates Charles Darwin, and wonders whether "dinosaurs were dumb."

Groombridge, Brian, ed. *Global Biodiversity: The Status of the Earth's Living Resources.* London; New York: Chapman & Hall, 1992. 585 pages. ISBN: 0412472406.

A massive review of biodiversity around the world, looking at the status of key ecosystems, the human uses of plants and animals, rates of change in natural systems, and so on.

Hopkin, Stephen P. *Biology of the Springtails.* Oxford; New York: Oxford University Press, 1997. 330 pages. ISBN: 0198540841.

Although some of the chapters will probably appeal only to the academic, many are filled with such fascinating facts as to appeal to a general reader.

Jones, Steve. *The Language of the Genes.* New York: Doubleday, 1994. 272 pages. ISBN: 0385473729.

Entertaining yet restrained, the celebrated geneticist gives a reliable account of what we do and do not know about our genes.

Jones, Steve, Robert Martin, and David Pilbeam, eds. *The Cambridge Encyclopedia of Human Evolution.* Cambridge [England]; New York: Cambridge University Press, 1992. 506 pages. ISBN: 0521323703.

Edited by a team led by Steve Jones, the well-known geneticist. A comprehensive and up-to-date survey of all aspects of human evolution, with some fascinating information.

Joyce, Christopher. *Earthly Goods.* Boston: Little, Brown, 1994. 304 pages. ISBN: 0316474088.

Tackles the complex relationships among medicinal plant hunters, local peoples, conservationists, and pharmaceutical companies, as well as emotive issues including "ownership"of biodiversity. The book is leavened with descriptions of eccentric scientists, the author's travels, and examples of useful plants.

Kamin, Leon. *The Science and Politics of IQ.* Potomac, MD: L. Erlbaum Associates; dist. by Halsted Press, New York, 1974. 183 pages. ISBN:0470455748.

More than twenty years old, this is a passionate and compelling study of the dubious idea that intelligence is determined by genetics.

Kimbrell, Andrew. *The Human Body Shop.* [San Francisco, CA]: Harper San Francisco, 1993. 348 pages. ISBN: 0062505246.

The author reviews the technological and commercial controls of human reproduction, arguing that our current legal and technical framework is inadequate to deal with advances in biotechnology.

King, Barry, ed. *Cell Biology.* London; Boston: Allen & Unwin, 1986. 265 pages. ISBN: 0045740461.

Different contributors give their accounts of some of the most interesting aspects of cell biology, including evolution, motility, and protein synthesis.

King, Robert C. *A Dictionary of Genetics.* 5th ed. New York: Oxford University Press, 1997. 439 pages. ISBN: 0195094425.

A-to-Z coverage of the terms used in the field of genetics.

Konner, Melvin. *The Tangled Wing.* New York: Holt, Rinehart, and Winston, 1982. 543 pages. ISBN: 003057062X.

A poetic and masterly survey of the biological constraints on human nature, including what animal behavior can and cannot tell us about ourselves

Lappé, Marc. *The Body's Edge: Our Cultural Obsession with the Skin.* New York: H. Holt, 1996. 242 pages. ISBN: 0805042083.

Very wide-ranging and accessible book that includes chapters on occupational illnesses, permeability, and animal skin, as well as anatomy and physiology.

Leakey, Richard. *The Origin of Humankind.* New York: BasicBooks, 1994. 171 pages. ISBN: 0465031358.

A brief and readable personal view of human fossils by today's leading authority on them.

Lear, Linda. *Rachel Carson: Witness for Nature.* New York: H. Holt, 1997. 634 pages. ISBN: 0805034277.

An exact and thorough account of Carson's life and writings.

Lewin, Roger. *Principles of Human Evolution.* Malden, MA: Blackwell Science, 1998. 526 pages. ISBN: 0865425426.

An up-to-date work that is pleasant to read despite being structured as a modular textbook.

Lewontin, Richard. *Biology as Ideology: The Doctrine of DNA.* New York: HarperPerennial, 1992. 128 pages. ISBN: 0060975199.

Lectures originally given on Canadian radio. A sustained attack on the idea that DNA controls people—or even cells.

Lorenz, Konrad. *King Solomon's Ring.* New York: New American Library, 1991. 202 pages. ISBN: 0451132297.

A gripping introduction to modern animal behavior by one of its founders, an advocate of the view that most animal behavior is genetically fixed or innate. Full of original insights and a pleasure to read.

Lovelock, James. *Gaia: A New Look at Life on Earth.* Oxford [Oxfordshire]; New York: Oxford University Press, 1987. 157 pages, includes bibliography. ISBN: 0192860305.

Proposes the still heretical theory that the planet itself is, biologically speaking, a self-regulating mechanism. Whether one agrees with Lovelock or not, he presents a sober and well-argued case, and his book helped focus attention on the role that different components play in the ecology of the planet, thus paving the way for discussion on biodiversity.

Maynard Smith, John, *The Theory of Evolution.* Cambridge [England]; New York: Cambridge University Press, 1993. 354 pages, includes index and bibliography. ISBN:0521451280.

Clear, authoritative, and readable account of the modern neo-Darwinian theory.

McFarland, David, ed. *The Oxford Companion to Animal Behavior.* Oxford; New York: Oxford University Press, 1982, 1981. 657 pages, includes index and bibliography. ISBN: 0198661207.

A standard handbook that anyone interested in animals can enjoy. Contains more than two hundred articles, each written by a specialist, on a wide range of topics.

McNeely, Jeffrey A., et al. *Conserving the World's Biological Diversity.* Washington, DC: World Resources Institute: Conservation International: World Wildlife Fund/U.S.: World Bank, 1990. 193 pages. ISBN: 0915825422.

The most comprehensive attempt so far to set an agenda for international conservation of biodiversity. Much of the thinking was later incorporated into proposals for the Convention on Biological Diversity.

Myers, Norman. *The Sinking Ark.* Oxford; New York: Pergamon Press, 1979. 307 pages.

Argues that we face a massive loss of species, particularly through tropical deforestation, and that conservation efforts are manifestly failing to counter this trend. In this book he also puts the practical case for preserving species, in terms of their use as genetic resources for foodstuffs, medicines, and so on; a theme which he has continued to developed in a number of further titles, including *The Primary Source.*

Nelkin, Dorothy, and M. Susan Lindee. *The DNA Mystique.* New York: Freeman, 1995. 276 pages. ISBN: 0716727099.

Two American authors give fascinating insights into the way the gene has invaded popular culture, turning up in comics, pop songs, and soap commercials.

Norse, Elliott A., ed. *Global Marine Biological Diversity.* Washington, DC: Island Press, 1993. 383 pages. ISBN: 1559632550.

Looks at the ecology of marine systems and at steps to address current problems facing the world's seas and oceans.

Noss, Reed F., and Allen Y. Cooperrider. *Saving Nature's Legacy.* Washington, DC: Island Press, 1994. 416 pages.

Guidelines on practical biodiversity conservation at a basic level.

Paul, William, ed. *Immunology.* New York: W. H. Freeman, 1991. 168 pages. ISBN: 0716722232.

Authoritative papers from *Scientific American* that cover the fascinating science of immunology. Difficult in parts but a wonderful book for conveying how clever cells can be.

Potter, Stephen, and Laurens Sargent. *Pedigree: Words from Nature.* New York: Taplinger Pub. Co., 1974. 320 pages. ISBN: 0800862481.

A wonderful collection of revelations about the origins of the names we give to plants and animals. From where, for example, does the porcupine get its name? And did you know that the collective noun for the starling is a *murmuration?*

Rachels, James. *Created from the Animals: The Moral Implications of Darwinism.* Oxford [England]; New York: Oxford University Press, 1991. 245 pages. ISBN: 0192861298.

A powerful defense of Darwin arguing that, correctly understood, Darwinism provides a strong theoretical basis for animal rights. Maintains that Darwin, in holding that humans are "created from the animals" laid the foundations for a new inclusivist ethic. Chapter 5 on "Morality without the Idea That Humans Are Special" is provocative and challenging. Written in an accessible style with wit and elegance.

Raven, Peter, Ray Evert, and Susan Eichorn. *Biology of Plants.* New York: Worth Publishers, 1992. 5th ed. 791 pages. ISBN: 0879015322.

As well as explaining form and function, the authors address the themes of heredity, evolutionary relationships, and ecology, providing a rounded picture of plant biology.

Reaka-Kudla, Marjorie, ed. *Biodiversity II.* Washington, DC: Joseph Henry Press, 1997. 551 pages. ISBN: 0309052270.

An interesting and enjoyable essay collection that provides much food for thought on the earth's biodiversity.

Ridley, Mark. *The Problems of Evolution.* Oxford [England]; New York: Oxford University Press, 1985. 159 pages, includes index and bibliography. ISBN: 0192191942.

Witty and cultivated essays on particular topics of controversy in the field of evolution.

Ridley, Matt. *The Red Queen.* New York: Maxwell Macmillan International, 1994. 405 pages. ISBN: 0026033402.

A brilliant look at the evolution of sex and its implications for behavior, providing a rich collection of insights into the private lives of a whole host of creatures.

Ritvo, Harriet. *The Platypus and the Mermaid and Other Figments of the Classifying Imagination.* Cambridge, MA.: Harvard University Press, 1997. 288 pages. ISBN: 0674673573.

A wide-ranging look at classification at its broadest. It is both authoritative and compelling.

Rose, Steven. *The Chemistry of Life.* Hammondsworth [England]; New York: Penguin, 1979. 301 pages. ISBN: 0140207902.

This is an up-to-date and authoritative survey of the chemical goings-on inside a cell. Rose is a professor at the Open University and here makes a technical subject reasonably approachable.

Rose, Steven, Leon Kamin, and Richard Lewontin. *Not in Our Genes.* New York: Pantheon Books, 1984. 322 pages. ISBN: 0394508173.

Political as well as biological, these essays are an argument against the so-called reductionist ideas that suggest genes are in charge.

Schrödinger, Erwin. *What Is Life?* Cambridge [England]: Cambridge University Press, 1992. 184 pages. ISBN: 0521427088.

This is one of the great science classics of the twentieth century, elegantly written by an eminent physicist. His theoretical approach to the question of what constitutes life involved an analysis of how genes must work, an analysis that set the agenda for the new postwar study of molecular biology.

Shiva, Vandana. *Biodiversity: Social and Ecological Consequences.* London; Atlantic Highlands, NJ: Zed Books, 1991. 123 pages. ISBN: 185649053X.

Proposes that biodiversity conservation has to be tackled by first addressing problems of intensive farming and North–South relations; one of a series of books by a noted Indian scientist and feminist; *Staying Alive,* by the same author, focuses on women's issues.

Tudge, Colin. *The Engineer in the Garden.* New York: Hill and Wang, 1995. 388 pages. ISBN: 0809042592.

Sometimes disturbing account of how animals and plants may grow up to be different as a result of genetic engineering.

Vroon, Piet. *Smell: The Secret Seducer.* New York: Farrar, Straus & Giroux, 1997. 226 pages. ISBN: 0374257043.

An exploration of the biology of the olfactory system and its importance in human affairs.

Warnock, Mary. *A Question of Life.* Oxford, UK; New York: B. Blackwell, 1985. 110 pages. ISBN: 0631142576.

The author is a British philosopher who has had enormous influence on the issue of what it is permissible to do with embryos. Intended for a broad general readership, this is a clear and balanced report of a committee she chaired that looked into human fertilization, embryology, and the ethics of assisted reproduction.

Watson, James D. *The Double Helix.* New York: Norton, 1980. 298 pages. ISBN: 039301245X.

A terrific firsthand account of perhaps the greatest scientific breakthroughs of our age—the discovery of the structure of DNA by the two young Cambridge scholars James Watson and Francis Crick.

Weiner, Jonathan. *The Beak of the Finch.* New York: Knopf; dist. by Random House, 1994. 332 pages. ISBN: 0679400036.

Written by a journalist, this books succeeds well in conveying the fascination inherent in the profoundly important long-term studies by Peter Grant and his coworkers into the action of natural selection on Galapagos finch populations.

Wilkie, Tom. *Perilous Knowledge.* Berkeley: University of California Press, 1993. 195 pages. ISBN: 0520085531.

Useful account of the Human Genome Project—the mapping and decoding of each and every one of our genes. Wilkie is a science journalist and well placed to understand and express public concerns over the moral consequences of molecular biology.

Williams, George C. *Adaptation and Natural Selection.* Princeton, NJ: Princeton University Press, 1966. 307 pages.

A seminal work that succeeds in combining inspiration with tough-minded correction of error. The book has exerted an increasing influence in the decades since it was written, and Williams is now respected as perhaps the dominant figure among American Darwinians.

Wilson, Edward O. *Sociobiology.* Cambridge, MA: Belknap Press of Harvard University Press, 1975. 697 pages. ISBN: 0674816218.

This groundbreaking study of the biological basis of the behavior of social animals was the subject of much controversy when it first appeared, principally because of the final chapter focusing on humans. Authoritative, encyclopedic, and readable.

Wilson, Edward O., ed. *Biodiversity.* Washington, DC: National Academy Press, 1988. 521 pages. ISBN: 0309037832.

The first book to crystallize the debate about biodiversity, which also brought the word itself to public attention. A fascinating array of articles, papers, and even poems about biological diversity around the world. Still the single best introduction to the subject. See also *The Diversity of Life* by Wilson himself for a readable introduction to the subject.

Wolpert, Lewis. *The Triumph of the Embryo.* Oxford [England]; New York: Oxford University Press, 1991. 211 pages. ISBN: 0198542437.

A limpid and engaging account of embryology and development—the coordinated process leading from a fertilized egg to an adult that is itself capable of becoming a parent.

6 World Wide Web Sites

Access Excellence
http://www.gene.com/ae/

U.S.-based site for biology teachers, sponsored by a biotechnology company, which has plenty to interest the casual browser as well—particularly its "What's New?" section, with weekly science reports and interviews with scientists making the news, and "About biotech"—an in-depth look at the field of biotechnology.

Amino Acids
http://www.chemie.fu-berlin.de/chemistry/bio/amino-acids_en.html

Small but interesting site giving the names and chemical structures of all the amino acids. The information is available in both English and German.

Angiosperm Anatomy
http://www.botany.uwc.ac.za:80/sci_ed/std8/anatomy/

Good general guide to angiosperms. The differences between monocotyledons and dicotyledons are set out here. The functions of roots, stems, leaves, and flowers are explained by readily understandable text and good accompanying diagrams.

Biology Timeline
http://www.zoologie.biologie.de/history.html

Chronology of important developments in the biological sciences. It includes items from the mentioning of hand pollination of date palms in 1800 B.C. to the Nobel Prize award for the discovery of site-directed mutagenesis in 1993. The site also includes a list of sources.

Breaking the Genetic Code

http://www.nih.gov/od/museum/neir1.htm

Museum exhibit Web site describing the Nobel Prize–winning work of Marshall W. Nirenberg on genetics. The site is broken down into three small pages with links to descriptions of the instruments used in Nirenberg's work. Along with the brief text description of the discovery several photographs and other images are also featured.

Bugs in the News!

http://falcon.cc.ukans.edu/~jbrown/bugs.html

Lively articles fill the user in not just on the microorganisms in the news, but also on immunity, antibiotics, and molecular-biology issues in the "real" world.

Carnivorous Plants FAQ

http://www.indirect.com/www/bazza/cps/faq/faq.html

Well-written source of general information about carnivorous plants. Each genus is presented with good text and pictures. There is advice on planting, growing, and feeding carnivorous plants, plus information on efforts to conserve endangered species.

Cells Alive

http://www.cellsalive.com/

Lively and attractive collection of microscopic and computer-generated images of living cells and microorganisms. It includes sections on HIV infection, penicillin, and how antibodies are made.

Francis Harry Compton Crick

http://kroeber.anthro.mankato.msus.edu/bio/francis_crick.htm

Profile of the life and achievements of the pioneer molecular biologist. It traces his upbringing and education and how he brought his knowledge of X-ray diffraction to his work with James Watson in unraveling the structure of DNA. There is a photo, a listing of Crick's major books and articles, and a bibliography.

Richard Dawkins

http://www.spacelab.net/~catalj/

Biographical information about Richard Dawkins, plus quotes, interviews, papers, articles, and excerpts from his books. This is an unofficial site gathering together a whole host of regularly updated information from a wide variety of sources.

Dennis Kunkel's Microscopy
http://www.pbrc.hawaii.edu/~kunkel/

This site is a photomicographer's dream—full of pictures taken with both light and electron microscopes. As well as several differing galleries of images, there is also information about microscopy and how the pictures were taken.

Dictionary of Cell Biology
http://www.mblab.gla.ac.uk/~julian/Dict.html

Searchable database of more than five thousand terms frequently encountered in reading modern biology literature. The dictionary can be searched as a whole, or in subsections such as "Disease," "Cytoskeleton," and "Nucleus, Genes, and DNA."

Theodosius Dobzhansky
http://kroeber.anthro.mankato.msus.edu/bio/Dobzhansky.htm

Profile of the pioneering geneticist. It traces his childhood interest in insects, the frustration of his ambitions in the Soviet Union, and his subsequent research in the United States. Includes photographs of Dobzhansky and a bibliography.

Dr. Frankenstein, I Presume?
http://www.salonmagazine.com/feb97/news/news2970224.html

Interview with the man who made the first cloned mammal, Dolly. Embryologist Dr. Ian Wilmut speaks to Andrew Ross about his worries and his future projects, about the distinction between science fiction and human cloning, and about what could go wrong with researching this delicate area.

Gertrude B. Elion Profile
http://www.achievement.org/autodoc/page/eli0pro-1

Description of the life and work of the Nobel Prize–winner Gertrude Elion. In addition to a biographical profile, the Web site contains a lengthy 1991 interview with Elion, along with a large number of photographs, video sequences, and audio clips.

Evolution: Theory and History
http://www.ucmp.berkeley.edu/history/evolution.html

Dedicated to the study of the history and theories associated with evolution, this site explores topics on classification, taxonomy, and dinosaur discoveries, and then looks at the key figures in the field and reviews their contributions.

The Evolutionist
http://www.lse.ac.uk/depts/cpnss/evolutionist/

Online magazine devoted to evolutionary ideas expressed in features, interviews, and comment. It currently includes an article on the recent political interest in Darwinism and a column on the limits of evolutionary theory.

Fern Resource Hub
http://www.inetworld.net/~sdfern/

Huge source of fern-related information for the fern hobbyist. Organized by the San Diego Fern Society, this is a clearinghouse of information for fern societies across the world. The site includes general information on ferns and how to grow them, lists of additional resources, an area where users can email experts with their questions on ferns, and news of upcoming fern events all over the world.

Fungi
http://www.herb.lsa.umich.edu/kidpage/factindx.htm

University-run network of hyperlinked pages on fungi from their earliest fossil records to their current ecology and life cycles, from how they are classified systematically to how they are studied. Although this page is not shy of technical terms, there are clear explanations and pictures to help the uninitiated.

Sir Frederick Gowland Hopkins
http://web.calstatela.edu/faculty/nthomas/hopkins.htm

Maintained by Nigel Thomas at California State University, Los Angeles, this page is devoted to the life and scientific work of Frederick Hopkins. Biographical information includes text and a time line of important moments in Hopkins's life. A special "science section" provides information on essential amino acids and vitamins. In addition to a bibliography of books by and about Hopkins, the site also offers a bibliography of texts about vitamins.

Human Anatomy Online
http://www.innerbody.com/indexbody.html

Fun, interactive, and educational site on the human body. The site is divided into many informative sections, including hundreds of images and animations.

Human Genome Project Information
http://www.ornl.gov/TechResources/Human_Genome/home.html

U.S.-based site devoted to this mammoth project—with news, progress reports, a molecular genetics primer, and links to other relevant sites.

Introduction to Proteins
http://biotech.chem.indiana.edu/pages/protein_intro.html

Introduction to the world of proteins with discussions of their structures and sequences, as well as descriptions of some major kinds of proteins. The site also includes a useful glossary and offers a variety of photographic shots.

Journey into Phylogenetic Systematics
http://www.ucmp.berkeley.edu/clad/clad4.html

Online exhibition about evolutionary theory with a specific emphasis on Phylogenetic Systematics: the way that biologists reconstruct the pattern of events that has led to the distribution and diversity of life. The site provides an introduction to the philosophy, methodology, and implication of cladistic analysis, with a separate section on the need for cladistics. Many of the scientific terms are included in an illuminating glossary.

Live a Life Page
http://alife.fusebox.com/

Fascinating page that brings together interactive programs to simulate ecological and evolutionary processes of evolution. Included is an adaptation of Richard Dawkins's Biomorphs program, described in *The Blind Watchmaker,* which enables the user to select "morphs" for certain qualities and watch as their offspring evolve.

Microbe Zoo

http://commtechlab.msu.edu/sites/dlc-me/zoo/

Colorful and interactive zoo of some of the microbes that surround us. It includes sections on the "domestic" microbes, the vampire ones that suck the life from other bacteria, the killers that destroy stone buildings, those in aquatic environments, and those that are to be found in beer, bread, chocolate, wine, and other consumables.

Micropropagation at Kew

http://www.rbgkew.org.uk/ksheets/microprop.html

Report of the work of the Micropropagation Unit at Britain's prestigious Kew Gardens. The site contains descriptions of how endangered species are cultured in vitro from seed or vegetative material. There are details of Kew's online publication, *Micropropagation News.*

Molecular Expressions: The Amino Acid Collection

http://micro.magnet.fsu.edu/aminoacids/index.html

Fascinating collection of images showing what all the known amino acids look like when photographed through a microscope. There is also a detailed article about the different amino acids.

Molecular Expressions: The DNA Collection

http://micro.magnet.fsu.edu/micro/gallery/dna/dna4.html

Spectacular gallery of DNA photographic representations in the laboratory as well as in vivo. This site also has links to several other sites offering photographs through a microscope of various substances, including computer chips and various pharmaceutical substances.

Molecular Expressions: The Vitamin Collection

http://micro.magnet.fsu.edu/vitamins/index.html

Fascinating collection of images showing what all the known vitamins look like when recrystallized and photographed through a microscope. There is also a brief article about vitamins.

Mystery of Smell

http://www.hhmi.org/senses/d/d110.htm

As part of a much larger site called "Seeing, Hearing, and Smelling the World," here is a page examining the way our sense of smell works. It is divided into four sections called "The Vivid World of Odors," "Finding the Odorant Receptors," "How Rats and Mice—and Probably Humans—Recognize Odors," and "The Memory of Smells." This site makes good use of images and animations to help with the explanations, so it is best viewed with an up-to-date browser.

Natural History of Genetics
http://raven.umnh.utah.edu/

Through a combination of scientific experts and teachers, this site offers an accessible and well-designed introduction to genetics. It includes several guided projects with experiments and explanations aimed initially at young teenage children. However, this site also includes "intermediate" and "expert" sections that will appeal to a wide variety of ages and levels of expertise. In addition to the experiments, the site includes sections on such topics as "core genetics," "teacher workshops," and "fun stuff."

Photosynthesis Directory
http://esg-www.mit.edu:8001/esgbio/ps/psdir.html

Wealth of scientific information concerning photosynthesis, its stages and its importance, from MIT in Boston. The site discusses issues such as the evolution and discovery of photosynthesis, the chloroplast, chlorophyll, and all steps of the light and dark reactions that take place during photosynthesis. The site also offers detailed diagrams of the procedures discussed.

Rachel Carson Homestead
http://www.rachelcarson.org/

Information on the life and legacy of the pioneering ecologist from the trust preserving Carson's childhood home. There is a biography of Carson, descriptions of books by and about Carson, and full details of the work of the conservation organizations continuing her work.

Roche-HIV Web Site
http://www.roche-hiv.com/

HIV virus Web site focusing on the properties of the virus and its treatment rather than the illness it causes. Roche say the purpose of the Web site is to "facilitate a better understanding of the virus and treatment issues, specifically resistance and adherence to HIV therapy." To this end they have included a multimedia presentation on the HIV lifecycle among other educational resources.

Royal Botanic Gardens, Kew
http://www.rbgkew.org.uk/

Kew Gardens' home page, with general visitor information, a history of the gardens, a guide to the main plant collections, and a searchable database.

Royal Horticultural Society (RHS)
http://www.rhs.org.uk/

Large source of information on horticulture in the United Kingdom. The contents include a guide to RHS gardens, the society's scientific and educational work, RHS publications, and information about the annual Chelsea Flower Show and other events.

Seeds of Life
http://www.vol.it/mirror/SeedsOfLife/home.html

Wealth of information about seeds and fruits, with information about the basic structure of a seed, fruit types, how seeds are dispersed, and seeds and humans, plus a mystery seed contest.

Short Botanical Glossary
http://www.anbg.gov.au/glossary/croft.html

Useful glossary of botanical terms. If you are confused about abaxial, zygote, or anything in between, this glossary will set you straight. A useful resource for high school botany students.

Survey of the Plant Kingdoms
http://www.mancol.edu/science/biology/plants_new/intro/start.html

Systematic guide to the nonanimal kingdoms (plants, fungi, protista, and monera)—their major groups, classification, and anatomy.

Tree of Life
http://phylogeny.arizona.edu/tree/phylogeny.html

Project designed to present information about the phylogenetic relationships and characteristics of organisms, illustrating the diversity and unity of living organisms.

Virtual Body
http://www.medtropolis.com/vbody/

If ever something was worth taking the time to download the Shockwave plug-in for, this is it. Authoritative and interactive anatomical animations complete with voice-overs guide you around the whole body, with sections on the brain, digestive system, heart, and skeleton.

Visible Embryo
http://visembryo.ucsf.edu/

Learn about the first four weeks of human development.

Visible Human Project
http://www.nlm.nih.gov/research/visible/visible_gallery.html

Sample images from a long-term U.S. project to collect a complete set of anatomically detailed, three-dimensional representations of the human body.

James D. Watson Profile
http://www.achievement.org/autodoc/page/wat0pro-1

Description of the life and works of the discoverer of the DNA molecule, James Watson. Watson won the Nobel Prize for his most famous discovery in 1962. The Web site contains not only a profile and biographical information, but also a lengthy 1991 interview with Watson accompanied by a large number of photographs, video sequences, and audio clips.

Innovations in Biology
Part II

7

Dictionary of Terms and Concepts

abdomen

In vertebrates, the part of the body below the thorax, containing the digestive organs; in insects and other arthropods, it is the hind part of the body. In mammals, the abdomen is separated from the thorax by the diaphragm, a sheet of muscular tissue; in arthropods, commonly by a narrow constriction. In mammals, the female reproductive organs are in the abdomen. In insects and spiders, it is characterized by the absence of limbs.

abiotic factor

A nonorganic variable within the ecosystem, affecting the life of organisms. Examples include temperature, light, and soil structure. Abiotic factors can be harmful to the environment, as when sulfur dioxide emissions from power stations produce acid rain.

abscissin, or abscissic acid

Plant hormone found in all higher plants. It is involved in the process of abscission and also inhibits stem elongation, germination of seeds, and the sprouting of buds.

abscission

In botany, the controlled separation of part of a plant from the main plant body—most commonly, the falling of leaves or the dropping of fruit controlled by abscissin. In deciduous plants the leaves are shed before the winter or dry season, whereas evergreen plants drop their leaves continually throughout the year. Fruitdrop, the abscission of fruit while still immature, is a naturally occurring process.

absorption

The taking up of one substance by another, such as a liquid by a solid (ink by blotting paper) or a gas by a liquid (ammonia by water).

abzyme

In biotechnology, an artificially created antibody that can be used like an enzyme to accelerate reactions.

acclimation, or acclimatization

The physiological changes induced in an organism by exposure to new environmental conditions. When humans move to higher altitudes, for example, the number of red blood cells rises to increase the oxygen-carrying capacity of the blood in order to compensate for the lower levels of oxygen in the air.

accommodation

The ability of the eye to focus on near or far objects by changing the shape of the lens.

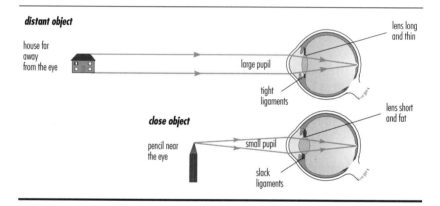

achene

Dry, one-seeded fruit that develops from a single ovary and does not split open to disperse the seed. Achenes commonly occur in groups—for example, the fruiting heads of buttercup *Ranunculus* and clematis. The outer surface may be smooth, spiny, ribbed, or tuberculate, depending on the species.

Achilles tendon

Tendon at the back of the ankle attaching the calf muscles to the heel bone. It is one of the largest tendons in the human body and can resist great tensional strain, but is sometimes ruptured by contraction of the muscles in sudden extension of the foot.

acid rain

Acidic precipitation thought to be caused principally by the release into the atmosphere of sulfur dioxide (SO_2) and oxides of nitrogen, which dissolve in pure rainwater making it acidic. Sulfur dioxide is formed by the burning of fossil fuels, such as coal, that contain high quantities of sulfur; nitrogen oxides are contributed from various industrial activities and from car exhaust fumes.

acquired character
Feature of the body that develops during the lifetime of an individual, usually as a result of repeated use or disuse, such as the enlarged muscles of a weightlifter.

action potential
A change in the potential difference (voltage) across the membrane of a nerve cell when an impulse passes along it. A change in potential (from about –60 to +45 millivolts) accompanies the passage of sodium and potassium ions across the membrane.

active transportation
In cells, the use of energy to move substances, usually molecules or ions, across a membrane.

adaptation
Derived from the Latin *adaptare,* "to fit to," adaptation is any change in the structure or function of an organism that allows it to survive and reproduce more effectively in its environment. In evolution, adaptation is thought to occur as a result of random variation in the genetic makeup of organisms coupled with natural selection. Species become extinct when they are no longer adapted to their environment—for instance, if the climate suddenly becomes colder.

adaptive radiation
In evolution, the formation of several species, with adaptations to different ways of life, from a single ancestral type. Adaptive radiation is likely to occur whenever members of a species migrate to a new habitat with unoccupied ecological niches. It is thought that the lack of competition in such niches allows sections of the migrant population to develop new adaptations, and eventually to become new species. The colonization of newly formed volcanic islands has led to the development of many unique species. The thirteen species of Darwin's finch on the Galápagos Islands, for example, are probably descended from a single species from the South American mainland. The parent stock evolved into different species that now occupy a range of diverse niches.

adenoids
Masses of lymphoid tissue, similar to tonsils, located in the upper part of the throat, behind the nose. They are part of a child's natural defenses against the entry of germs but usually shrink and disappear by the age of ten.

adenosine diphosphate (ADP)
The chemical product formed in cells when adenosine triphosphate (ATP) breaks down to release energy.

adenosine triphosphate (ATP)
Nucleotide molecule found in all cells. It can yield large amounts of energy and is used to drive the thousands of biological processes needed to sustain life,

growth, movement, and reproduction. Green plants use light energy to manufacture ATP as part of the process of photosynthesis. In animals, ATP is formed by the breakdown of glucose molecules, usually obtained from the carbohydrate component of a diet, in a series of reactions termed respiration. It is the driving force behind muscle contraction and the synthesis of complex molecules needed by individual cells.

ADH
See **antidiuretic hormone (ADH)**, part of the system maintaining a correct salt/water balance in vertebrates.

adipose tissue
Type of connective tissue of vertebrates that serves as an energy reserve, and also pads some organs. It is commonly called fat tissue, and consists of large spherical cells filled with fat. In mammals, major layers are in the inner layer of skin and around the kidneys and heart.

adolescence
In the human life cycle, the period between the beginning of puberty and adulthood.

ADP
See **adenosine diphosphate (ADP)**.

adrenal glands, or suprarenal glands
Two triangular glands situated on top of the kidney. The adrenals are soft and yellow and consist of two parts: the cortex and medulla. The **cortex** (outer part) secretes various steroid hormones and other hormones that control salt and water metabolism and regulate the use of carbohydrates, proteins, and fats. The **medulla** (inner part) secretes the hormones adrenaline and noradrenaline which, during times of stress, cause the heart to beat faster and harder, increase blood flow to the heart and muscle cells, and dilate airways in the lungs, thereby delivering more oxygen to cells throughout the body and in general preparing the body for "fight or flight."

adrenaline, or epinephrine
Hormone secreted by the medulla of the adrenal glands. Adrenaline is synthesized from a closely related substance, noradrenaline, and the two hormones are released into the bloodstream in situations of fear or stress.

aerenchyma
Plant tissue with numerous air-filled spaces between the cells. It occurs in the stems and roots of many aquatic plants where it aids buoyancy and transportation of oxygen around the plant.

aerobic
Term used to describe those organisms that require oxygen (usually dissolved in water) for the efficient release of energy contained in food molecules, such as

glucose. They include almost all organisms (plants as well as animals) with the exception of certain bacteria, yeasts, and internal parasites.

afterbirth
In mammals, the placenta, umbilical cord, and ruptured membranes, which become detached from the uterus and are expelled soon after birth.

after-ripening
Dormancy undergone by the seeds of some plants before germination can occur but after the seed has completed development. The length of the after-ripening period in different species may vary from a few weeks to many months. It helps seeds to germinate at a time when conditions are most favorable for growth. In some cases the embryo is not fully mature at the time of dispersal and must develop further before germination can take place. Other seeds do not germinate even when the embryo is mature, probably owing to growth inhibitors within the seed that must be leached out or broken down before germination can begin.

agar
Jellylike carbohydrate, obtained from seaweeds. It is used mainly in microbiological experiments as a culture medium for growing bacteria and other microorganisms. The agar is resistant to breakdown by microorganisms, remaining a solid jelly throughout the course of the experiment.

aggression
Behavior used to intimidate or injure another organism (of the same or of a different species), usually for the purposes of gaining territory, a mate, or food. Aggression often involves an escalating series of threats aimed at intimidating an opponent without having to engage in potentially dangerous physical contact. Aggressive signals include roaring by red deer, snarling by dogs, the fluffing-up of feathers by birds, and the raising of fins by some species of fish.

aging
In common usage, the period of deterioration of the physical condition of a living organism that leads to death; in biological terms, the entire life process.

agonist
A muscle that contracts and causes a movement. Contraction of an agonist is complemented by relaxation of its antagonist. For example, the biceps (in the front of the upper arm) bends the arm while the triceps (lying behind the biceps) straightens the arm. *See also* **antagonistic muscles.**

air passages
The nose, pharynx, larynx, trachea, and bronchi. When a breath is taken, air passes through high narrow passages on each side of the nose where it is warmed and moistened and particles of dust are removed. Food and air passages meet

and cross in the pharynx. The larynx lies in front of the lower part of the pharynx and it is the organ where the voice is produced using the vocal cords. The air passes the glottis (the opening between the vocal cords) and enters the trachea. The trachea leads into the chest and divides above the heart into two bronchi. The bronchi carry the air to the lungs and they subdivide to form a succession of fine tubes and, eventually, a network of capillaries that allows the exchange of gases between the inspired air and the blood.

air sac
In birds, a thin-walled extension of the lungs. There are nine of these and they extend into the abdomen and bones, effectively increasing lung capacity. In mammals, it is another name for the alveoli in the lungs, and in some insects, for widenings of the trachea.

albumin, or albumen
Any of a group of sulfur-containing proteins. The best known is in the form of egg white; others occur in milk, and as a major component of serum. They are soluble in water and dilute salt solutions and are coagulated by heat.

alimentary canal
In animals, the tube through which food passes; it extends from the mouth to the anus. It is a complex organ, adapted for digestion. In human adults, it is about 9 m/30 ft long, consisting of the mouth cavity, pharynx, esophagus, stomach, and the small and large intestines.

allele
One of two or more alternative forms of a gene at a given position (locus) on a chromosome, caused by a difference in the DNA. Blue and brown eyes in humans are determined by different alleles of the gene for eye color.

allometry
A regular relationship between a given feature (for example, the size of an organ) and the size of the body as a whole, when this relationship is not a simple proportion of body size. Thus, an organ may increase in size proportionately faster, or slower, than body size does. For example, a human baby's head is much larger in relation to its body than is an adult's.

alpha-interferon
See interferon.

alternation of generations
Typical life cycle of terrestrial plants and some seaweeds, in which there are two distinct forms occurring alternately: diploid (having two sets of chromosomes) and haploid (one set of chromosomes). The diploid generation produces haploid

spores by meiosis and is called the sporophyte, while the haploid generation produces gametes (sex cells) and is called the gametophyte. The gametes fuse to form
a diploid zygote, which develops into a new sporophyte; thus the sporophyte and
gametophyte alternate.

altruism

Helping another individual of the same species to reproduce more effectively, as a
direct result of which the altruist may leave fewer offspring itself. Female honey
bees (workers) behave altruistically by rearing sisters in order to help their mother,
the queen bee, reproduce, and forgo any possibility of reproducing themselves.

alveoli

See **alveolus.**

alveolus

Plural, one of the many thousands of tiny air sacs in the lungs in which exchange
of oxygen and carbon dioxide takes place between air and the bloodstream.

ameba

Plural *amebas,* one of the simplest living animals, consisting of a single cell and
belonging to the protozoa group. The body consists of colorless protoplasm. Its
activities are controlled by the nucleus, and it feeds by flowing around and en-
gulfing organic debris. It reproduces by binary fission. Some species of ameba are
harmful parasites.

amino acid

Water-soluble organic molecule, mainly composed of carbon, oxygen, hydrogen,
and nitrogen, containing both a basic amino group (NH_2) and an acidic carboxyl
($COOH$) group. They are small molecules able to pass through membranes. When

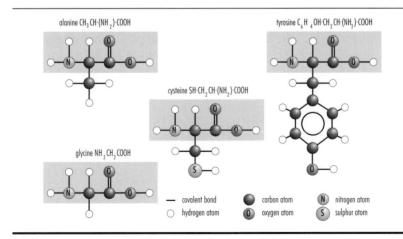

alanine $CH_3 \cdot CH \cdot (NH_2) \cdot COOH$

tyrosine $C_6H_4OH \cdot CH_2 \cdot CH \cdot (NH_2) \cdot COOH$

cysteine $SH \cdot CH_2 \cdot CH \cdot (NH_2) \cdot COOH$

glycine $NH_2 \cdot CH_2 \cdot COOH$

— covalent bond
○ hydrogen atom
● carbon atom
(O) oxygen atom
(N) nitrogen atom
(S) sulphur atom

two or more amino acids are joined together, they are known as peptides; proteins are made up of peptide chains folded or twisted in characteristic shapes.

amphibian
From Greek, meaning "double life," a member of the vertebrate class Amphibia, which generally spend their larval (tadpole) stage in fresh water, transferring to land at maturity (after metamorphosis) and generally returning to water to breed. Like fish and reptiles, they continue to grow throughout life, and cannot maintain a temperature greatly differing from that of their environment. The class contains 4,553 known species, 4,000 of which are frogs and toads, 390 salamanders, and 163 caecilians (wormlike in appearance).

amylase
One of a group of enzymes that break down starches into their component molecules (sugars) for use in the body. It occurs widely in both plants and animals. In humans, it is found in saliva and in pancreatic juices.

anabolism
Process of building up body tissue, promoted by the influence of certain hormones. It is the constructive side of metabolism, as opposed to catabolism.

anaerobic
Applies to organisms not requiring oxygen for the release of energy from food molecules such as glucose. Anaerobic organisms include many bacteria, yeasts, and internal parasites.

anal canal
See anus.

analgesic
Agent for relieving pain. Opiates alter the perception or appreciation of pain and are effective in controlling "deep" visceral (internal) pain. Nonopiates, such as aspirin, paracetamol, and NSAIDs (nonsteroidal anti-inflammatory drugs), relieve musculoskeletal pain and reduce inflammation in soft tissues.

analogous
Term describing a structure that has a similar function to a structure in another organism, but not a similar evolutionary path. For example, the wings of bees and of birds have the same purpose—to give powered flight—but have different origins. Compare *homologous*.

anatomy
Study of the structure of the body and its component parts, especially the human body, as distinguished from physiology, which is the study of bodily functions.

androecium

Male part of a flower, comprising a number of stamens.

androgen

General name for any male sex hormone, of which testosterone is the most important. They are all steroids and are principally involved in the production of male secondary sexual characteristics (such as beard growth).

anemophily

Type of pollination in which the pollen is carried on the wind. Anemophilous flowers are usually unscented, have either very reduced petals and sepals or lack them altogether, and do not produce nectar. In some species the flowers are borne in catkins. Male and female reproductive structures are commonly found in separate flowers. The male flowers have numerous exposed stamens, often on long filaments; the female flowers have long, often branched, feathery stigmas.

angiosperm

Flowering plant in which the seeds are enclosed within an ovary, which ripens into a fruit. Angiosperms are divided into monocotyledons (single seed leaf in the embryo) and dicotyledons (two seed leaves in the embryo). They include the majority of flowers, herbs, grasses, and trees, except conifers.

animals

From Latin *anima* meaning "breath" or "life," members of the kingdom Animalia, one of the major categories of living things, the science of which is zoology. Animals are all heterotrophs (they obtain their energy from organic substances produced by other organisms); they have eukaryotic cells (the genetic material is contained within a distinct nucleus) bounded by a thin cell membrane rather than the thick cell wall of plants. Most animals are capable of moving around for at least part of their life cycle. *See also* **metazoa.**

annelid

Any segmented worm of the phylum Annelida. Annelids include earthworms, leeches, and marine worms such as lugworms.

annual plant

Plant that completes its life cycle within one year, during which time it germinates, grows to maturity, bears flowers, produces seed, and dies.

annual rings, or growth rings

Concentric rings visible on the wood of a cut tree trunk or other woody stem. Each ring represents a period of growth when new xylem is laid down to replace tissue being converted into wood (secondary xylem). The wood formed from xylem produced in the spring and early summer has larger and more numerous vessels than the wood formed from xylem produced in autumn when growth is slowing down. The result is a clear boundary between the pale spring wood and the denser, darker autumn wood. Annual rings may be used to estimate the age of the plant.

anoxia, or hypoxia
Deprivation of oxygen, a condition that rapidly leads to collapse or death unless immediately reversed.

antagonist
A muscle that relaxes in response to the contraction of its agonist muscle. The biceps, in the front of the upper arm, bends the arm while the triceps, lying behind the biceps, straightens the arm. *See also* **antagonistic muscles.**

antagonistic muscles
In the body, a pair of muscles allowing coordinated movement of the skeletal joints. The extension of the arm, for example, requires one set of muscles to relax, while another set contracts. The individual components of antagonistic pairs can be classified into extensors (muscles that straighten a limb) and flexors (muscles that bend a limb).

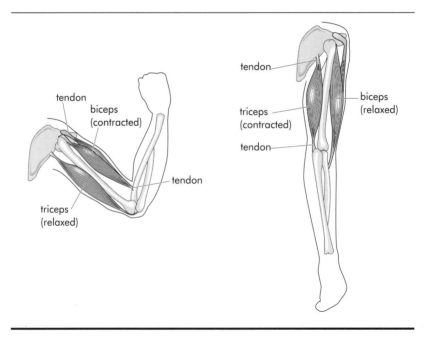

antenna
In zoology, an appendage ("feeler") on the head. Insects, centipedes, and millipedes each have one pair of antennae but there are two pairs in crustaceans, such

as shrimps. In insects, the antennae are involved with the senses of smell and touch; they are frequently complex structures with large surface areas that increase the ability to detect scents.

anther
In a flower, the terminal part of a stamen in which the pollen grains are produced. It is usually borne on a slender stalk or filament and has two lobes, each containing two chambers, or pollen sacs, within which the pollen is formed.

antheridium
Organ producing the male gametes, or antherozoids, in algae, bryophytes (mosses and liverworts), and pteridophytes (ferns, club mosses, and horsetails). It may be either single-celled, as in most algae, or multicellular, as in bryophytes and pteridophytes.

antherozoid
Motile (or independently moving) male gamete produced by algae, bryophytes (mosses and liverworts), pteridophytes (ferns, club mosses, and horsetails), and some gymnosperms (notably the cycads). Antherozoids are formed in an antheridium and, after being released, swim by means of one or more flagella, to the female gametes. Higher plants have nonmotile male gametes contained within pollen grains.

antibody
Protein molecule produced in the blood by lymphocytes in response to the presence of foreign or invading substances (antigens); such substances include the proteins carried on the surface of infecting microorganisms. Antibody production is only one aspect of immunity in vertebrates.

anticoagulant
Substance that inhibits the formation of blood clots. Common anticoagulants are heparin, produced by the liver and some white blood cells, and derivatives of coumarin. Anticoagulants are used medically in the prevention and treatment of thrombosis and heart attacks. Anticoagulant substances are also produced by blood-feeding animals, such as mosquitoes, leeches, and vampire bats, to keep the victim's blood flowing.

antidiuretic hormone (ADH)
Part of the system maintaining a correct salt/water balance in vertebrates.

antigen
Any substance, usually a protein, that causes the production of antibodies by the body's immune system. Common antigens include the proteins carried on the surface of bacteria, viruses, and pollen grains. The proteins of incompatible blood groups or tissues also act as antigens, which has to be taken into account in medical procedures such as blood transfusions and organ transplants.

anus, or anal canal
The opening at the end of the alimentary canal that allows undigested food and other waste materials to pass out of the body, in the form of feces. In humans, the term is also used to describe the last 4 cm/1.5 in of the alimentary canal. The anus is found in all types of multicellular animal except the coelenterates (sponges) and the platyhelminths (flatworms), which have a mouth only.

aorta
The body's main artery, arising from the left ventricle of the heart in birds and mammals. Carrying freshly oxygenated blood, it arches over the top of the heart and descends through the trunk, finally splitting in the lower abdomen to form the two iliac arteries. Loss of elasticity in the aorta provides evidence of atherosclerosis, which may lead to heart disease.

aposematic coloration
The technical name for warning coloration markings that make a dangerous, poisonous, or foul-tasting animal particularly conspicuous and recognizable to a predator. Examples include the yellow and black stripes of bees and wasps and the bright red or yellow colors of many poisonous frogs.

appendix
A short, blind-ended tube attached to the cecum. It has no known function in humans, but in herbivores it may be large, containing millions of bacteria that secrete enzymes to digest grass (as no vertebrate can secrete enzymes that will digest cellulose, the main constituent of plant cell walls).

aquarium
Tank or similar container used for the study and display of living aquatic plants and animals. The same name is used for institutions that exhibit aquatic life. The oceanarium or seaquarium is a large display for marine life forms.

aquatic
Living in water. All life on earth originated in the early oceans, because the aquatic environment has several advantages for organisms. Dehydration is almost impossible, temperatures usually remain stable, and the density of water provides physical support.

aqueous humor
Watery fluid found in the chamber between the cornea and lens of the vertebrate eye. Similar to blood serum in composition, it is constantly renewed.

arboretum
Collection of trees. An arboretum may contain a wide variety of species or just closely related species or varieties—for example, different types of pine tree.

Archaea

Group of microorganisms that are without a nucleus and have a single chromosome. All are strict anaerobes, that is, they are killed by oxygen. This is thought to be a primitive condition and to indicate that Archaea are related to the earliest life forms, which appeared about four billion years ago, when there was little oxygen in the earth's atmosphere. They are found in undersea vents, hot springs, the Dead Sea, and salt pans, and have even adapted to garbage dump sites.

archegonium

From Greek *arche* meaning "origin" and *gonos,* "offspring," female sex organ found in bryophytes (mosses and liverworts), pteridophytes (ferns, club mosses, and horsetails), and some gymnosperms. It is a multicellular, flask-shaped structure consisting of two parts: the swollen base or venter containing the egg cell, and the long, narrow neck. When the egg cell is mature, the cells of the neck dissolve, allowing the passage of the male gametes, or antherozoids.

aril

Accessory seed cover other than a fruit; it may be fleshy and sometimes brightly colored, woody, or hairy. In flowering plants (angiosperms), it is often derived from the stalk that originally attached the ovule to the ovary wall. Examples of arils include the bright-red, fleshy layer surrounding the yew seed (yews are gymnosperms so they lack true fruits) and the network of hard filaments that partially covers the nutmeg seed and yields the spice known as mace.

artery

Vessel that carries blood from the heart to the rest of the body. It is built to withstand considerable pressure, having thick walls that contain smooth muscle fibers. During contraction of the heart muscle, arteries expand in diameter to allow for the sudden increase in pressure that occurs; the resulting pulse or pressure wave can be felt at the wrist. Not all arteries carry oxygenated (oxygen-rich) blood; the pulmonary arteries convey deoxygenated (oxygen-poor) blood from the heart to the lungs.

arthropod

Member of the phylum Arthropoda; an invertebrate animal with jointed legs and a segmented body with a horny or chitinous casing (exoskeleton), which is shed periodically and replaced as the animal grows. Included are arachnids such as spiders and mites, as well as crustaceans, millipedes, centipedes, and insects.

artificial selection

Selective breeding of individuals that exhibit the particular characteristics that a plant or animal breeder wishes to develop. In plants, desirable features might include resistance to disease, high yield (in crop plants), or attractive appearance. In animal breeding, artificial selection has led to the development of particular breeds of cattle for improved meat production.

ascorbic acid

$C_6H_8O_6$ or vitamin C, a relatively simple organic acid found in citrus fruits and vegetables. It is soluble in water and destroyed by prolonged boiling, so soaking or overcooking of vegetables reduces their vitamin C content. Lack of ascorbic acid results in scurvy.

asexual reproduction

Reproduction that does not involve the manufacture and fusion of sex cells, nor the necessity for two parents. The process carries a clear advantage in that there is no need to search for a mate or to develop complex pollinating mechanisms; every asexual organism can reproduce on its own. Asexual reproduction can therefore lead to a rapid population buildup.

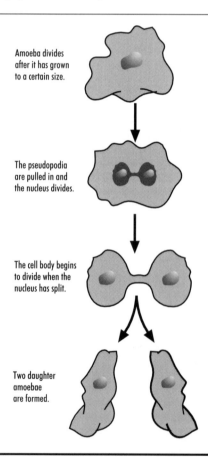

Amoeba divides after it has grown to a certain size.

The pseudopodia are pulled in and the nucleus divides.

The cell body begins to divide when the nucleus has split.

Two daughter amoebae are formed.

assimilation

In animals, the process by which absorbed food molecules, circulating in the blood, pass into the cells and are used for growth, tissue repair, and other metabolic activities. The actual destiny of each food molecule depends not only on its type, but also on the body's requirements at that time.

assortative mating
In population genetics, selective mating in a population between individuals that are genetically related or have similar characteristics. If sufficiently consistent, assortative mating can theoretically result in the evolution of new species without geographical isolation.

atavism
From Latin *atavus,* meaning "ancestor"; in genetics, the reappearance of a characteristic not apparent in the immediately preceding generations; in psychology, the manifestation of primitive forms of behavior.

ATP
See **adenosine triphosphate (ATP).**

atrium
Either of the two upper chambers of the heart. The left atrium receives freshly oxygenated blood from the lungs via the pulmonary vein; the right atrium receives deoxygenated blood from the vena cava. Atrium walls are thin and stretch easily to allow blood into the heart. On contraction, the atria force blood into the thick-walled ventricles, which then give a second, more powerful beat.

auditory canal
Tube leading from the outer ear opening to the eardrum. It is found only in animals whose eardrums are located inside the skull, principally mammals and birds.

autolysis
The destruction of a cell after its death by the action of its own enzymes, which break down its structural molecules.

autonomic nervous system
In mammals, the part of the nervous system that controls those functions not controlled voluntarily, including the heart rate, activity of the intestines, and the production of sweat. There are two divisions of the autonomic nervous system. The sympathetic system responds to stress, when it speeds the heart rate, increases blood pressure, and generally prepares the body for action. The parasympathetic system is more important when the body is at rest, since it slows the heart rate, decreases blood pressure, and stimulates the digestive system.

autoradiography
A technique for following the movement of molecules within an organism, especially a plant, by labeling with a radioactive isotope that can be traced on

photographs. It is used to study photosynthesis, where the pathway of radioactive carbon dioxide can be traced as it moves through the various chemical stages.

autosome

Any chromosome in the cell other than a sex chromosome. Autosomes are of the same number and kind in both males and females of a given species.

autotroph

Any living organism that synthesizes organic substances from inorganic molecules by using light or chemical energy. Autotrophs are the primary producers in all food chains since the materials they synthesize and store are the energy sources of all other organisms. All green plants and many planktonic organisms are autotrophs, using sunlight to convert carbon dioxide and water into sugars by photosynthesis.

auxin

Plant hormone that promotes stem and root growth in plants. Auxins influence many aspects of plant growth and development, including cell enlargement, inhibition of development of axillary buds, tropisms, and the initiation of roots. Synthetic auxins are used in rooting powders for cuttings, and in some weedkillers, where high auxin concentrations cause such rapid growth that the plants die. They are also used to prevent premature fruitdrop in orchards. The most common naturally occurring auxin is known as indoleacetic acid, or IAA. It is produced in the shoot apex and transported to other parts of the plant.

axil

Upper angle between a leaf (or bract) and the stem from which it grows. Organs developing in the axil, such as shoots and buds, are termed axillary, or lateral.

axon

Long threadlike extension of a nerve cell that conducts electrochemical impulses away from the cell body toward other nerve cells, or toward an effector organ such as a muscle. Axons terminate in synapses, junctions with other nerve cells, muscles, or glands.

B cell, or B lymphocyte

Immune cell that produces antibodies. Each B cell produces just one type of antibody, specific to a single antigen. Lymphocytes are related to T cells.

B lymphocyte

See B cell.

bacillus

Member of a group of rodlike bacteria that occur everywhere in the soil and air. Some are responsible for diseases such as anthrax or for causing food spoilage.

backcross

Breeding technique used to determine the genetic makeup of an individual organism.

bacteria

(Singular *bacterium*), microscopic single-celled organisms lacking a nucleus. Bacteria are widespread, present in soil, air, and water, and as parasites on and in other living things. Some parasitic bacteria cause disease by producing toxins, but others are harmless and may even benefit their hosts. Bacteria usually reproduce by binary fission (dividing into two equal parts), and this may occur approximately every twenty minutes. It is thought that only 1 to 10 percent of the world's bacteria have been identified.

bacteriology

The study of bacteria.

bacteriophage

Virus that attacks bacteria. Such viruses are now of use in genetic engineering.

balance of nature

In ecology, the idea that there is an inherent equilibrium in most ecosystems, with plants and animals interacting so as to produce a stable, continuing system of life on earth. The activities of human beings can, and frequently do, disrupt the balance of nature.

baldness

Loss of hair from the scalp, common in older men. Its onset and extent are influenced by genetic makeup and the level of male sex hormones. There is no cure, and expedients such as hair implants may have no lasting effect. Hair loss in both sexes may also occur as a result of ill health or radiation treatment, such as for cancer. Alopecia, a condition in which the hair falls out in patches, is different from the "male-pattern baldness" described above.

ball-and-socket joint

Joint allowing considerable movement in three dimensions, for instance the joint between the pelvis and the femur. To facilitate movement, such joints are rimmed with cartilage and lubricated by synovial fluid. The bones are kept in place by ligaments and moved by muscles.

bark

Protective outer layer on the stems and roots of woody plants, composed mainly of dead cells. To allow for expansion of the stem, the bark is continually added to from within, and the outer surface often becomes cracked or is shed as scales. Trees deposit a variety of chemicals in their bark, including poisons. Many of these chemical substances have economic value because they can be used in the manufacture of drugs. Quinine, derived from the bark of the *Cinchona* tree, is used to fight malarial infections; curare, an anesthetic used in medicine, comes from the *Strychnus toxifera* tree in the Amazonian rain forest.

baroreceptor

A specialized nerve ending that is sensitive to pressure. There are baroreceptors in various regions of the heart and circulatory system (carotid sinus, aortic arch, atria, pulmonary veins, and left ventricle). Increased pressure in these structures stimulates the baroreceptors, which relay information to the medulla, providing an important mechanism in the control of blood pressure.

basal metabolic rate (BMR)

Minimum amount of energy needed by the body to maintain life. It is measured when the subject is awake but resting and includes the energy required to keep the heart beating, sustain breathing, repair tissues, and keep the brain and nerves functioning. Measuring the subject's consumption of oxygen gives an accurate value for BMR, because oxygen is needed to release energy from food.

base pair

In biochemistry, the linkage of two base (purine or pyrimidine) molecules in DNA. They are found in nucleotides and form the basis of the genetic code.

basidiocarp

Spore-bearing body, or "fruiting body," of all basidiomycete fungi, except the rusts and smuts. A well-known example is the edible mushroom *Agaricus brunnescens*. Other types include globular basidiocarps (puffballs) or flat ones that project from tree trunks (brackets). They are made up of a mass of tightly packed, intermeshed hyphae.

beriberi

Nutritional disorder occurring mostly in the tropics and resulting from a deficiency of vitamin B_1 (thiamine). The disease takes two forms: in one edema (waterlogging of the tissues) occurs; in the other there is severe emaciation. There is nerve degeneration in both forms and many victims succumb to heart failure.

berry

Fleshy, many-seeded fruit that does not split open to release the seeds. The outer layer of tissue, the exocarp, forms an outer skin that is often brightly colored to attract birds to eat the fruit and thus disperse the seeds. Examples of berries are the tomato and the grape.

beta-interferon

See interferon.

bicuspid valve, or mitral valve

In the left side of the heart, a flap of tissue that prevents blood flowing back into the atrium when the ventricle contracts.

biennial plant

Plant that completes its life cycle in two years. During the first year it grows vegetatively and the surplus food produced is stored in its perennating organ, usually the root. In the following year these food reserves are used for the production of leaves, flowers, and seeds, after which the plant dies. Many root vegetables are biennials, including the carrot *Daucus carota* and parsnip *Pastinaca sativa.* Some garden plants that are grown as biennials are actually perennials, for example, the wallflower *Cheiranthus cheiri.*

bile

Brownish alkaline fluid produced by the liver. Bile is stored in the gall bladder and is intermittently released into the duodenum (small intestine) to aid digestion. Bile consists of bile salts, bile pigments, cholesterol, and lecithin. Bile salts assist in the breakdown and absorption of fats; bile pigments are the breakdown products of old red blood cells that are passed into the gut to be eliminated with the feces.

binary fission

A form of asexual reproduction, whereby a single-celled organism, such as the ameba, divides into two smaller "daughter" cells. It can also occur in a few simple multicellular organisms, such as sea anemones, producing two smaller sea anemones of equal size.

binomial system of nomenclature

The system in which all organisms are identified by a two-part Latinized name. Devised by the biologist Linnaeus, it is also known as the Linnaean system. The first name is capitalized and identifies the genus; the second identifies the species within that genus, for example, *Quercus alba,* the white oak.

biochemistry

Science concerned with the chemistry of living organisms: the structure and reactions of proteins (such as enzymes), nucleic acids, carbohydrates, and lipids.

biodegradable

Capable of being broken down by living organisms, principally bacteria and fungi. In biodegradable substances, such as food and sewage, the natural processes of decay lead to compaction and liquefaction, and to the release of nutrients that are then recycled by the ecosystem.

biodiversity

Contraction of the words *biological diversity,* measure of the variety of the earth's animal, plant, and microbial species; of genetic differences within species; and of the ecosystems that support those species. Its maintenance is important for ecological stability and as a resource for research into, for example, new drugs and crops. In the twentieth century, the destruction of habitats is believed to have resulted in the most severe and rapid loss of biodiversity in the history of the planet.

bioengineering

The application of engineering to biology and medicine. Common applications include the design and use of artificial limbs, joints, and organs, including hip joints and heart valves.

biogeography

Study of how and why plants and animals are distributed around the world, in the past as well as in the present; more specifically, a theory describing the geographical distribution of species developed by Canadian-born U.S. ecologist Robert MacArthur and U.S. zoologist Edward O. Wilson. The theory argues that for many species, ecological specializations mean that suitable habitats are patchy in their occurrence. Thus for a dragonfly, ponds in which to breed are separated by large tracts of land, and for edelweiss adapted to alpine peaks the deep valleys between cannot be colonized.

biological clock

Regular internal rhythm of activity, produced by unknown mechanisms and not dependent on external time signals. Such clocks are known to exist in almost all animals, and also in many plants, fungi, and unicellular organisms; the first biological clock gene in plants was isolated in 1995 by a team of U.S. researchers. In higher organisms, there appears to be a series of clocks of graded importance. For example, although body temperature and activity cycles in human beings are normally "set" to twenty-four hours, the two cycles may vary independently, showing that two clock mechanisms are involved.

biological control

Control of pests such as insects and fungi through biological means, rather than the use of chemicals. This can include breeding resistant crop strains; inducing sterility in the pest; infecting the pest species with disease organisms; or introducing the pest's natural predator. Biological control tends to be naturally self-regulating, but as ecosystems are complex, it is difficult to predict all the consequences of introducing a biological controlling agent.

biological oxygen demand (BOD)

Amount of dissolved oxygen taken up by microorganisms in a sample of water. Since these microorganisms live by decomposing organic matter, and the amount of oxygen used is proportional to their number and metabolic rate, BOD can be used as a measure of the extent to which the water is polluted with organic compounds.

biology

From the Greek *bios* "life," and *logos* "discourse," science of life. Biology includes all the life sciences—for example, anatomy and physiology (the study of the structure of living things), cytology (the study of cells), zoology (the study of animals) and botany (the study of plants), ecology (the study of habitats and the interaction of living species), animal behavior, embryology, taxonomy, and plant breeding. Increasingly in the twentieth century, biologists have concentrated on molecular structures: biochemistry, biophysics, and genetics (the study of inheritance and variation).

bioluminescence

Production of light by living organisms. It is a feature of many deep-sea fishes, crustaceans, and other marine animals. On land, bioluminescence is seen in some nocturnal insects such as glow-worms and fireflies, and in certain bacteria and fungi. Light is usually produced by the oxidation of luciferin, a reaction catalyzed by the enzyme luciferase. This reaction is unique, being the only known biological oxidation that does not produce heat. Animal luminescence is involved in communication, camouflage, or the luring of prey, but its function in other organisms is unclear.

biomass

The total mass of living organisms present in a given area. It may be specified for a particular species (such as earthworm biomass) or for a general category (such as herbivore biomass). Estimates also exist for the entire global plant biomass. Measurements of biomass can be used to study interactions between organisms, the stability of those interactions, and variations in population numbers. Where dry biomass is measured, the material is dried to remove all water before weighing.

biome

Broad natural assemblage of plants and animals shaped by common patterns of vegetation and climate. Examples include the tundra biome and the desert biome.

biometry

Literally, the measurement of living things, but generally used to mean the application of mathematics to biology. The term is now largely obsolete, since mathematical or statistical work is an integral part of most biological disciplines.

biophysics

Application of physical laws to the properties of living organisms. Examples include using the principles of mechanics to calculate the strength of bones and muscles, and thermodynamics to study plant and animal energetics.

biorhythm

Rhythmic change, mediated by hormones, in the physical state and activity patterns of certain plants and animals that have seasonal activities. Examples include winter hibernation, spring flowering or breeding, and periodic migration. The hormonal changes themselves are often a response to changes in day length (photoperiodism); they signal the time of year to the animal or plant. Other biorhythms are innate and continue even if external stimuli such as day length are removed. These include a twenty-four-hour or circadian rhythm, a twenty-eight-day or circalunar rhythm (corresponding to the phases of the moon), and even a year-long rhythm in some organisms.

biosensor
Device based on microelectronic circuits that can directly measure medically significant variables for the purpose of diagnosis or monitoring treatment. One such device measures the blood sugar level of diabetics using a single drop of blood, and shows the result on a liquid crystal display within a few minutes.

biosonar
See **echolocation**.

biosphere
The narrow zone that supports life on our planet. It is limited to the waters of the earth, a fraction of its crust, and the lower regions of the atmosphere. The biosphere is made up of all the earth's ecosystems. It is affected by external forces such as the sun's rays, which provide energy, the gravitational effects of the sun and moon, and cosmic radiations.

biosynthesis
Synthesis of organic chemicals from simple inorganic ones by living cells—for example, the conversion of carbon dioxide and water to glucose by plants during photosynthesis. Other biosynthetic reactions produce cell constituents including proteins and fats.

biotechnology
Industrial use of living organisms to manufacture food, drugs, or other products. The brewing and baking industries have long relied on the yeast microorganism for fermentation purposes, while the dairy industry employs a range of bacteria and fungi to convert milk into cheeses and yogurts. Enzymes, whether extracted from cells or produced artificially, are central to most biotechnological applications.

biotin, or vitamin H
Vitamin of the B complex, found in many different kinds of food; egg yolk, liver, legumes, and yeast contain large amounts. Biotin is essential to the metabolism of fats. Its absence from the diet may lead to dermatitis.

bird
Backboned animal of the class Aves, the biggest group of land vertebrates (nearly 8,500 species), characterized by warm blood, feathers, wings, breathing through lungs, and egg-laying by the female. Birds are bipedal; feet are usually adapted for perching and never have more than four toes. Hearing and eyesight are well developed, but the sense of smell is usually poor. No existing species of bird possesses teeth. Most birds fly, but some groups (such as ostriches) are flightless, and

others include flightless members. Many communicate by sounds (nearly half of all known species are songbirds) or by visual displays, in connection with which many species are brightly colored, usually the males. Birds have highly developed patterns of instinctive behavior.

birth

Act of producing live young from within the body of female animals. Both viviparous and ovoviviparous animals give birth to young. In viviparous animals, embryos obtain nourishment from the mother via a placenta or other means. In ovoviviparous animals, fertilized eggs develop and hatch in the oviduct of the mother and gain little or no nourishment from maternal tissues.

bivalent

A name given to the pair of homologous chromosomes during reduction division (meiosis). In chemistry, the term is sometimes used to describe an element or group with a valence of two, although the term *divalent* is more common.

bladder

Hollow elastic-walled organ that stores the urine produced in the kidneys. It is present in the urinary systems of some fishes, most amphibians, some reptiles, and all mammals. Urine enters the bladder through two ureters, one leading from each kidney, and leaves it through the urethra.

blastocyst

In mammals, the hollow ball of cells that is an early stage in the development of the embryo, roughly equivalent to the blastula of other animal groups.

blastomere

A cell formed in the first stages of embryonic development, after the splitting of the fertilized ovum, but before the formation of the blastula or blastocyst.

blastula

Early stage in the development of a fertilized egg, when the egg changes from a solid mass of cells (the morula) to a hollow ball of cells (the blastula), containing a fluid-filled cavity (the blastocoel).

blight

Any of a number of plant diseases caused mainly by parasitic species of fungus, which produce a whitish appearance on leaf and stem surfaces; for example, potato blight *(Phytophthora infestans)*. General damage caused by aphids or pollution is sometimes known as blight.

blind spot

Area where the optic nerve and blood vessels pass through the retina of the eye. No visual image can be formed as there are no light-sensitive cells in this part of the retina.

blood

Fluid circulating in the arteries, veins, and capillaries of vertebrate animals; the term also refers to the corresponding fluid in those invertebrates that possess a closed circulatory system. Blood carries nutrients and oxygen to each body cell and removes waste products, such as carbon dioxide. It is also important in the immune response and, in many animals, in the distribution of heat throughout the body.

blood clotting

Complex series of events (known as the blood clotting cascade) that prevents excessive bleeding after injury. It is triggered by vitamin K. The result is the formation of a meshwork of protein fibers (fibrin) and trapped blood cells over the cut blood vessels.

blood group

Any of the types into which blood is classified according to the presence or absence of certain antigens on the surface of its red cells. Red blood cells of one individual may carry molecules on their surface that act as antigens in another individual whose red blood cells lack these molecules. The two main antigens are designated A and B. These give rise to four blood groups: having A only (A), having B only (B), having both (AB), and having neither (O). Each of these groups may or may not contain the rhesus factor. Correct typing of blood groups is vital in transfusion, since incompatible types of donor and recipient blood will result in coagulation, with possible death of the recipient.

blood vessel

Tube that conducts blood either away from or toward the heart in multicellular animals. The principal types are arteries, which conduct blood away from the heart; veins, which conduct blood toward the heart; and capillaries, which conduct blood from arteries to veins. Arteries always carry oxygenated blood and veins deoxygenated blood, with the exception of the pulmonary artery, which carries deoxygenated blood from the heart to the lungs, and the pulmonary vein, which carries oxygenated blood from the lungs to the heart.

bloom

Whitish powdery or waxlike coating over the surface of certain fruits that easily rubs off when handled. It often contains yeasts that live on the sugars in the fruit. The term *bloom* is also used to describe a rapid increase in number of certain species of algae found in lakes, ponds, and oceans.

blubber

Thick layer of fat under the skin of marine mammals, which provides an energy store and an effective insulating layer, preventing the loss of body heat to the surrounding water. Blubber has been used (when boiled down) in engineering, food processing, cosmetics, and printing, but all of these products can now be produced synthetically.

blue-green algae, or cyanobacteria

Single-celled, primitive organisms that resemble bacteria in their internal cell organization, sometimes joined together in colonies or filaments. Blue-green algae are among the oldest known living organisms and, with bacteria, belong to the kingdom Monera; remains have been found in rocks up to 3.5 billion years old. They are widely distributed in aquatic habitats, on the damp surfaces of rocks and trees, and in the soil.

BMR

See basal metabolic rate (BMR).

bone

Hard connective tissue constituting the skeleton of most vertebrate animals. Bone is composed of a network of collagen fibers impregnated with mineral salts (largely calcium phosphate and calcium carbonate), a combination that gives it great density and strength, comparable in some cases with that of reinforced concrete. Enclosed within this solid matrix are bone cells, blood vessels, and nerves. The interior of the long bones of the limbs consists of a spongy matrix filled with a soft marrow that produces blood cells.

botanical garden

Place where a wide range of plants is grown, providing the opportunity to see a botanical diversity not likely to be encountered naturally.

botany

From Greek *botane*, "herb," the study of living and fossil plants, including form, function, interaction with the environment, and classification.

bovine somatotropin (BST)

Hormone that increases an injected cow's milk yield by 10 to 40 percent. It is a protein naturally occurring in milk and breaks down within the human digestive tract into harmless amino acids. However, doubts have arisen recently as to whether such a degree of protein addition could in the long term be guaranteed harmless either to cattle or to humans.

brachiopod, or lamp shell

Any member of the phylum Brachiopoda, marine invertebrates with two shells, resembling but totally unrelated to bivalves. There are about three hundred living species; they were much more numerous in past geological ages. They are suspension feeders, ingesting minute food particles from water. In brachiopods, a single internal organ, the lophophore, handles feeding, aspiration, and excretion.

bract

Leaflike structure in whose axil a flower or inflorescence develops. Bracts are generally green and smaller than the true leaves. However, in some plants they may be brightly colored and conspicuous, taking over the role of attracting pollinating insects to the flowers, whose own petals are small; examples include the poinsettia *Euphorbia pulcherrima* and bougainvillea.

brain

In higher animals, a mass of interconnected nerve cells forming the anterior part of the central nervous system, whose activities it coordinates and controls. In vertebrates, the brain is contained by the skull. At the base of the brainstem, the medulla oblongata contains centers for the control of respiration, heartbeat rate and strength, and blood pressure. Overlying this is the cerebellum, which is concerned with coordinating complex muscular processes such as maintaining posture and moving limbs. The cerebral hemispheres (cerebrum) are paired outgrowths of the front end of the forebrain, in early vertebrates mainly con-

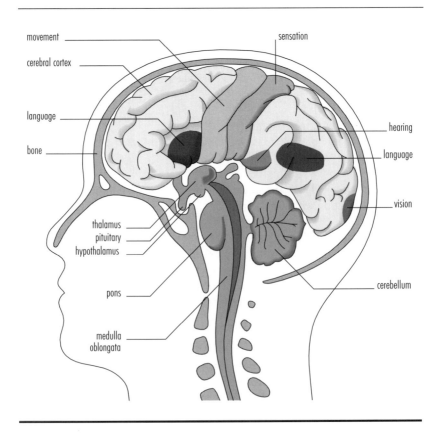

cerned with the senses, but in higher vertebrates greatly developed and involved in the integration of all sensory input and motor output, and in thought, emotions, memory, and behavior.

brainstem
Region where the top of the spinal cord merges with the undersurface of the brain, consisting largely of the medulla oblongata and midbrain.

breast
One of a pair of organs on the chest of the human female, also known as a mammary gland. Each of the two breasts contains milk-producing cells and a network of tubes or ducts that lead to openings in the nipple.

breathing
In terrestrial animals, the muscular movements whereby air is taken into the lungs and then expelled, a form of gas exchange. Breathing is sometimes referred to as external respiration, for true respiration is a cellular (internal) process.

breathing rate
The number of times a minute the lungs inhale and exhale. The rate increases during exercise because the muscles require an increased supply of oxygen and nutrients. At the same time very active muscles produce a greater volume of carbon dioxide, a waste gas that must be removed by the lungs via the blood.

breed
Recognizable group of domestic animals, within a species, with distinctive characteristics that have been produced by artificial selection.

breeding
The crossing and selection of animals and plants to change the characteristics of an existing breed or cultivar (variety), or to produce a new one.

broad-leaved tree
Another name for a tree belonging to the angiosperms, such as ash, beech, oak, maple, or birch. The leaves are generally broad and flat, in contrast to the needle-like leaves of most conifers.

bronchiole
Small-bore air tube found in the vertebrate lung responsible for delivering air to the main respiratory surfaces. Bronchioles lead off from the larger bronchus and branch extensively before terminating in the many thousand alveoli that form the bulk of lung tissue.

bronchus

One of a pair of large tubes (bronchi) branching off from the windpipe and passing into the vertebrate lung. Apart from their size, bronchi differ from the bronchioles in possessing cartilaginous rings, which give rigidity and prevent collapse during breathing movements.

bryophyte

Member of the Bryophyta, a division of the plant kingdom containing three classes: the Hepaticae (liverwort), Musci (moss), and Anthocerotae (hornwort). Bryophytes are generally small, low-growing, terrestrial plants with no vascular (water-conducting) system as in higher plants. Their life cycle shows a marked alternation of generations. Bryophytes chiefly occur in damp habitats and require water for the dispersal of the male gametes (antherozoids).

bud

Undeveloped shoot usually enclosed by protective scales; inside is a very short stem and numerous undeveloped leaves, or flower parts, or both. Terminal buds are found at the tips of shoots, while axillary buds develop in the axils of the leaves, often remaining dormant unless the terminal bud is removed or damaged. Adventitious buds may be produced anywhere on the plant, their formation sometimes stimulated by an injury, such as that caused by pruning.

budding

Type of asexual reproduction in which an outgrowth develops from a cell to form a new individual. Most yeasts reproduce in this way.

bulb

Underground bud with fleshy leaves containing a reserve food supply and with roots growing from its base. Bulbs function in vegetative reproduction and are characteristic of many monocotyledonous plants such as the daffodil, snowdrop, and onion.

bulbil

Small bulb that develops above ground from a bud. Bulbils may be formed on the stem from axillary buds, as in members of the saxifrage family, or in the place of flowers, as seen in many species of onion *Allium*. They drop off the parent plant and develop into new individuals, providing a means of vegetative reproduction and dispersal.

bur, or burr

In botany, a type of "false fruit" or pseudocarp, surrounded by numerous hooks; for instance, that of burdock *Arctium*, where the hooks are formed from bracts surrounding the flowerhead. Burs catch in the feathers or fur of passing animals, and thus may be dispersed over considerable distances.

byssus

Tough protein fibers secreted by the foot of sessile (fixed) bivalves, such as mussels, as a means of attachment to rocks.

caecum, or cecum

In the digestive system of animals, a blind-ending tube branching off from the first part of the large intestine, terminating in the appendix. It has no function in humans but is used for the digestion of cellulose by some grass-eating mammals.

callus

In botany, a tissue that forms at a damaged plant surface. Composed of large, thin-walled parenchyma cells, it grows over and around the wound, eventually covering the exposed area.

In animals, a callus is a thickened pad of skin, formed where there is repeated rubbing against a hard surface. In humans, calluses often develop on the hands and feet of those involved in heavy manual work.

calyptra

In mosses and liverworts, a layer of cells that encloses and protects the young sporophyte (spore capsule), forming a sheathlike hood around the capsule. The term is also used to describe the root cap, a layer of parenchyma cells covering the end of a root that gives protection to the root tip as it grows through the soil. This is constantly being worn away and replaced by new cells from a special meristem, the calyptrogen.

calyx

Collective term for the sepals of a flower, forming the outermost whorl of the perianth. It surrounds the other flower parts and protects them while in bud. In some flowers, for example, the campions *Silene,* the sepals are fused along their sides, forming a tubular calyx.

cambium

In botany, a layer of actively dividing cells (lateral meristem) found within stems and roots, that gives rise to secondary growth in perennial plants, causing an increase in girth. There are two main types of cambium: vascular cambium, which gives rise to secondary xylem and phloem tissues, and cork cambium (or phellogen), which gives rise to secondary cortex and cork tissues.

camouflage

Colors or structures that allow an animal to blend with its surroundings to avoid detection by other animals. Camouflage can take the form of matching the background color, of countershading (darker on top, lighter below, to counteract natural shadows), or of irregular patterns that break up the outline of the animal's body. More elaborate camouflage involves closely resembling a feature of the natural environment, as with the stick insect; this is closely akin to mimicry.

Campylobacter

Genus of bacteria that cause serious outbreaks of gastroenteritis. They grow best at 43°C/109.4°F, and so are well suited to the digestive tract of birds. Poultry is therefore the most likely source of a *Campylobacter* outbreak, although the bacteria can also be transmitted via beef or milk. *Campylobacter* can survive in water for up to fifteen days, so may be present in drinking water if supplies are contaminated by sewage or reservoirs are polluted by sea gulls.

canine

In mammalian carnivores, any of the long, often pointed teeth found at the front of the mouth between the incisors and premolars. Canine teeth are used for catching prey, for killing, and for tearing flesh. They are absent in herbivores such as rabbits and sheep, and are much reduced in humans.

capillarity

Spontaneous movement of liquids up or down narrow tubes, or capillaries. The movement is due to unbalanced molecular attraction at the boundary between the liquid and the tube. If liquid molecules near the boundary are more strongly attracted to molecules in the material of the tube than to other nearby liquid molecules, the liquid will rise in the tube. If liquid molecules are less attracted to the material of the tube than to other liquid molecules, the liquid will fall.

capillary

Narrowest blood vessel in vertebrates, 0.008–0.02 mm/0.00031496–0.000787402 inches in diameter, barely wider than a red blood cell. Capillaries are distributed as beds, complex networks connecting arteries and veins. Capillary walls are extremely thin, consisting of a single layer of cells, and so nutrients, dissolved gases, and waste products can easily pass through them. This makes the capillaries the main area of exchange between the fluid (lymph) bathing body tissues and the blood. They provide a large surface area in order to maximize diffusion.

capitulum

In botany, a flattened or rounded head (inflorescence) of numerous, small, stalkless flowers. The capitulum is surrounded by a circlet of petal-like bracts and has the appearance of a large, single flower.

capsule

In botany, a dry, usually many-seeded fruit formed from an ovary composed of two or more fused carpels, which splits open to release the seeds. The same term is used for the spore-containing structure of mosses and liverworts; this is borne at the top of a long stalk or seta.

carapace

Protective covering of many animals, particularly the arched bony plate characteristic of the order Chelonia (tortoises, terrapins, and turtles), and the shield that protects the fore parts of crustaceans, such as crabs.

carbon cycle

Sequence by which carbon circulates and is recycled through the natural world. Carbon dioxide is released into the atmosphere by living things as a result of respiration. The CO_2 is taken up and converted into carbohydrates during photosynthesis by plants and by organisms such as diatoms and dinoflagellates in the oceanic plankton; the oxygen component is released back into the atmosphere. The carbon they accumulate is later released back into circulation in various ways. The simplest occurs when an animal eats a plant and carbon is transferred from, say, a leaf cell to the animal body. Carbon is also released through the decomposition of decaying plant matter, and the burning of fossil fuels such as coal (fossilized plants). The oceans absorb 25 to 40 percent of all carbon dioxide released into the atmosphere.

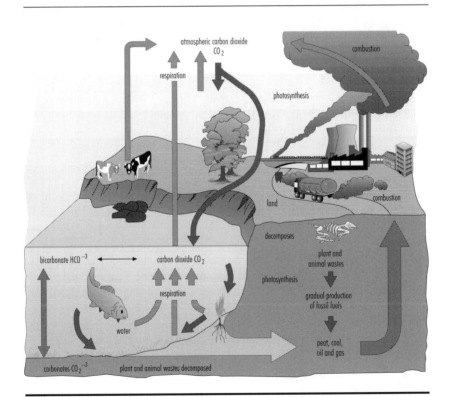

carbon dioxide (CO_2)

Colorless, odorless gas, slightly soluble in water and denser than air. It is formed by the complete oxidation of carbon.

carboxylic acid

See **fatty acid.**

carcinogen

Any agent that increases the chance of a cell's becoming cancerous, including various chemical compounds, some viruses, X-rays, and other forms of ionizing radiation. The term is often used more narrowly to mean chemical carcinogens only.

carcinoma

Malignant tumor arising from the skin, the glandular tissues, or the mucous membranes that line the gut and lungs.

carnassial tooth

One of a powerful scissorlike pair of molars, found in all mammalian carnivores except seals. Carnassials are formed from an upper premolar and lower molar and are shaped to produce a sharp cutting surface. Carnivores such as dogs transfer meat to the back of the mouth, where the carnassials slice up the food ready for swallowing.

carnivore

In zoology, mammal of the order Carnivora. Although its name describes the flesh-eating ancestry of the order, it includes pandas, which are herbivorous, and civet cats, which eat fruit.

carotene

Naturally occurring pigment of the carotenoid group. Carotenes produce the orange, yellow, and red colors of carrots, tomatoes, oranges, and crustaceans.

carotenoid

Any of a group of yellow, orange, red, or brown pigments found in many living organisms, particularly in the chloroplasts of plants. There are two main types, the carotenes and the xanthophylls. Both types are long-chain lipids (fats).

carotid artery

One of a pair of major blood vessels, one on each side of the neck, supplying blood to the head.

carpel

Female reproductive unit in flowering plants (angiosperms). It usually comprises an ovary containing one or more ovules, the stalk or style, and a stigma at its top, which receives the pollen. A flower may have one or more carpels, and they may be separate or fused together. Collectively the carpels of a flower are known as the gynoecium.

carrying capacity
In ecology, the maximum number of animals of a given species that a particular area can support. When the carrying capacity is exceeded, there is insufficient food (or other resources) for the members of the population. The population may then be reduced by emigration, reproductive failure, or death through starvation.

cartilage
Flexible bluish-white connective tissue made up of the protein collagen. In cartilaginous fish it forms the skeleton; in other vertebrates it forms the greater part of the embryonic skeleton and is replaced by bone in the course of development, except in areas of wear such as bone endings, and the discs between the backbones. It also forms structural tissue in the larynx, nose, and external ear of mammals.

caryopsis
Dry, one-seeded fruit in which the wall of the seed becomes fused to the carpel wall during its development. It is a type of achene, and therefore develops from one ovary and does not split open to release the seed. Caryopses are typical of members of the grass family (Gramineae), including the cereals.

casein
Main protein of milk, from which it can be separated by the action of acid, the enzyme rennin, or bacteria (souring); it is also the main protein in cheese. Casein is used as a protein supplement in the treatment of malnutrition. It is used commercially in cosmetics, glues, and as a sizing for coating paper.

catabolism
The destructive part of metabolism where living tissue is changed into energy and waste products. It is the opposite of anabolism. It occurs continuously in the body, but is accelerated during many disease processes, such as fever, and in starvation.

catalyst
Substance that alters the speed of, or makes possible, a chemical or biochemical reaction but remains unchanged at the end of the reaction. Enzymes are natural biochemical catalysts. In practice most catalysts are used to speed up reactions.

catecholamine
Chemical that functions as a neurotransmitter or a hormone. Dopamine, adrenaline (epinephrine), and noradrenaline (norepinephrine) are catecholamines.

catkin

In flowering plants (angiosperms), a pendulous inflorescence, bearing numerous small, usually unisexual flowers. The tiny flowers are stalkless and the petals and sepals are usually absent or much reduced in size. Many types of trees bear catkins, including willows, poplars, and birches. Most plants with catkins are wind-pollinated, so the male catkins produce large quantities of pollen. Some gymnosperms also have catkinlike structures that produce pollen, for example, the swamp cypress *Taxodium*.

cell

The basic structural unit of life. It is the smallest unit capable of independent existence that can reproduce itself exactly. All living organisms—with the exception of viruses—are composed of one or more cells. Single-cell organisms such as bacteria, protozoa, and other microorganisms are termed unicellular, while plants and animals, which contain many cells, are termed multicellular organisms. Highly complex organisms such as human beings consist of billions of cells, all of which are adapted to carry out specific functions—for instance, groups of these specialized cells are organized into tissues and organs. Although these cells may differ widely in size, appearance, and function, their essential features are similar. Cells divide by mitosis, or by meiosis when gametes are being formed.

cell differentiation

In developing embryos, the process by which cells acquire their specialization, such as heart cells, muscle cells, skin cells, and brain cells. The seven-day-old human preembryo consists of thousands of individual cells, each of which is destined to assist in the formation of individual organs in the body.

cell division

The process by which a cell divides, either meiosis, associated with sexual reproduction, or mitosis, associated with growth, cell replacement, or repair. Both forms involve the duplication of DNA and the splitting of the nucleus.

cell membrane, or plasma membrane

Thin layer of protein and fat surrounding cells that controls substances passing between the cytoplasm and the intercellular space. The cell membrane is semipermeable, allowing some substances to pass through and some not.

cell sap

Dilute fluid found in the large central vacuole of many plant cells. It is made up of water, amino acids, glucose, and salts. The sap has many functions, including storage of useful materials, and provides mechanical support for nonwoody plants.

cellulose

Complex carbohydrate composed of long chains of glucose units, joined by chemical bonds called glycosidic links. It is the principal constituent of the cell wall of

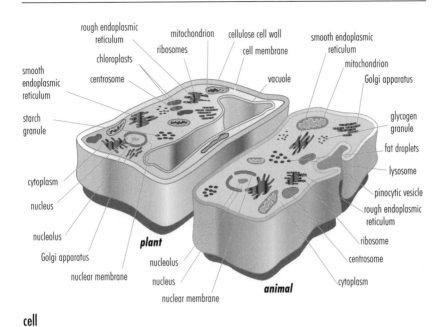

cell

higher plants, and a vital ingredient in the diet of many herbivores. Molecules of cellulose are organized into long, unbranched microfibrils that give support to the cell wall. No mammal produces the enzyme cellulase, necessary for digesting cellulose; mammals such as rabbits and cows are only able to digest grass because the bacteria present in their gut can manufacture it.

cell wall

In plants, the tough outer surface of the cell. It is constructed from a mesh of cellulose and is very strong and relatively inelastic. Most living cells are turgid (swollen with water) and develop an internal hydrostatic pressure (wall pressure) that acts against the cellulose wall. The result of this turgor pressure is to give the cell, and therefore the plant, rigidity. Plants that are not woody are particularly reliant on this form of support.

central dogma

In genetics and evolution, the fundamental belief that genes can affect the nature of the physical body, but that changes in the body (acquired character, for example, through use or accident) cannot be translated into changes in the genes.

central nervous system (CNS)

The brain and spinal cord, as distinct from other components of the nervous system. The CNS integrates all nervous function.

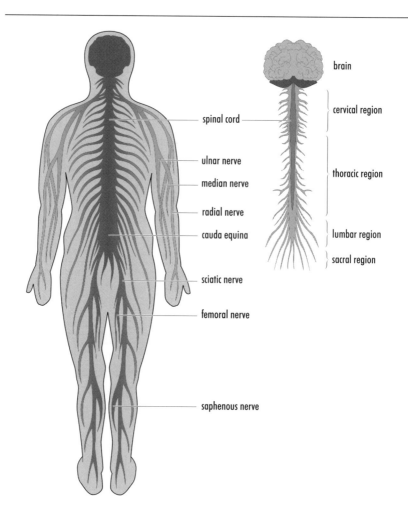

central nervous system

centriole

Structure found in the cells of animals that plays a role in the processes of meiosis and mitosis (cell division).

cephalopod

Any predatory marine mollusk of the class Cephalopoda, with the mouth and head surrounded by tentacles. Cephalopods are the most intelligent, the fastest-moving, and the largest of all animals without backbones, and there are remarkable luminescent forms that swim or drift at great depths. They have the most highly developed nervous and sensory systems of all invertebrates, the eye in some closely paralleling that found in vertebrates. Examples include squid, octopus, and cuttlefish. Shells are rudimentary or absent in most cephalopods.

cereal
Grass grown for its edible, nutrient-rich, starchy seeds. The term refers primarily to wheat, oats, rye, and barley, but may also refer to corn, millet, and rice. Cereals contain about 75 percent complex carbohydrates and 10 percent protein, plus fats and fiber (roughage). They store well. If all the world's cereal crop were consumed as whole-grain products directly by humans, everyone could obtain adequate protein and carbohydrates; however, a large proportion of cereal production in affluent nations is used as animal feed to boost the production of meat, dairy products, and eggs.

cerebellum
Part of the brain of vertebrate animals that controls muscle tone, movement, balance, and coordination. It is relatively small in lower animals such as newts and lizards, but large in birds, since flight demands precise coordination. The human cerebellum is also well developed, because of the need for balance when walking or running, and for finely coordinated hand movements.

cerebrum
Part of the vertebrate brain, formed from the two paired cerebral hemispheres, separated by a central fissure. In birds and mammals it is the largest and most developed part of the brain. It is covered with an infolded layer of gray matter, the cerebral cortex, which integrates brain functions. The cerebrum coordinates all voluntary activity.

chalaza
Glutinous mass of transparent albumen supporting the yolk inside birds' eggs. The chalaza is formed as the egg slowly passes down the oviduct, when it also acquires its coiled structure.

chemosynthesis
Method of making protoplasm (contents of a cell) using the energy from chemical reactions, in contrast to the use of light energy employed for the same purpose in photosynthesis. The process is used by certain bacteria, which can synthesize organic compounds from carbon dioxide and water using the energy from special methods of respiration.

chemotaxis
The property that certain cells have of attracting or repelling other cells. For example, white blood cells are attracted to the site of infection by the release of substances during certain types of immune response.

chemotropism

Movement by part of a plant in response to a chemical stimulus. The response by the plant is termed "positive" if the growth is toward the stimulus or "negative" if the growth is away from the stimulus.

childbirth

The expulsion of a baby from its mother's body following pregnancy. In a broader sense, it is the period of time involving labor and delivery of the baby.

chimera, or chimaera

An organism composed of tissues that are genetically different. Chimeras can develop naturally if a mutation occurs in a cell of a developing embryo, but are more commonly produced artificially by implanting cells from one organism into the embryo of another.

chitin

Complex long-chain compound, or polymer; a nitrogenous derivative of glucose. Chitin is widely found in invertebrates. It forms the exoskeleton of insects and other arthropods. It combines with protein to form a covering that can be hard and tough, as in beetles, or soft and flexible, as in caterpillars and other insect larvae. It is insoluble in water and resistant to acids, alkalis, and many organic solvents. In crustaceans such as crabs, it is impregnated with calcium carbonate for extra strength.

chlamydia

Viruslike bacteria that live parasitically in animal cells and cause disease in humans and birds. Chlamydiae are thought to be descendants of bacteria that have lost certain metabolic processes. In humans, a strain of chlamydia causes trachoma, a disease found mainly in the tropics (a leading cause of blindness); venereally transmitted chlamydiae cause genital and urinary infections.

chlorophyll

Green pigment present in most plants; it is responsible for the absorption of light energy during photosynthesis. The pigment absorbs the red and blue-violet parts of sunlight but reflects the green, thus giving plants their characteristic color.

chloroplast

Structure (organelle) within a plant cell containing the green pigment chlorophyll. Chloroplasts occur in most cells of the green plant that are exposed to light, often in large numbers. Typically, they are flattened and disclike, with a double membrane enclosing the stroma, a gel-like matrix. Within the stroma are stacks of fluid-containing cavities, or vesicles, where photosynthesis occurs.

chlorosis

Abnormal condition of green plants in which the stems and leaves turn pale green or yellow. The yellowing is due to a reduction in the levels of the green chlorophyll pigments. It may be caused by a deficiency in essential elements (such as magnesium, iron, or manganese), a lack of light, genetic factors, or viral infection.

cholecalciferol, or vitamin D

Fat-soluble chemical important in the uptake of calcium and phosphorous for bones. It is found in liver, fish oils, and margarine. It can be produced in the skin, provided that the skin is adequately exposed to sunlight. Lack of vitamin D leads to rickets and other bone diseases.

chordate

Animal belonging to the phylum Chordata, which includes vertebrates, sea squirts, amphioxi, and others. All these animals, at some stage of their lives, have a supporting rod of tissue (notochord or backbone) running down their bodies.

chorion

Outermost of the three membranes enclosing the embryo of reptiles, birds, and mammals; the amnion is the innermost membrane.

choroid

Layer found at the rear of the eye beyond the retina. By absorbing light that has already passed through the retina, it stops back-reflection and so prevents blurred vision.

chromosome

Structure in a cell nucleus that carries the genes. Each chromosome consists of one very long strand of DNA, coiled and folded to produce a compact body. The point on a chromosome where a particular gene occurs is known as its locus. Most higher organisms have two copies of each chromosome, together known as a homologous pair (they are diploid) but some have only one (they are haploid). There are forty-six chromosomes in a normal human cell.

chyme

General term for the stomach contents. Chyme resembles a thick creamy fluid and is made up of partly digested food, hydrochloric acid, and a range of enzymes.

cilia

Singular *cilium*, small hairlike organs on the surface of some cells, particularly the cells lining the upper respiratory tract. Their wavelike movements waft particles of dust and debris toward the exterior. Some single-celled organisms move by means of cilia. In multicellular animals, they keep lubricated surfaces clear of debris. They also move food in the digestive tracts of some invertebrates.

ciliary muscle

Ring of muscle surrounding and controlling the lens inside the vertebrate eye, used in accommodation (focusing). Suspensory ligaments, resembling spokes of a wheel, connect the lens to the ciliary muscle and pull the lens into a flatter shape when the muscle relaxes. When the muscle is relaxed the lens has its longest focal length and focuses rays from distant objects. On contraction, the lens returns to its normal spherical state and therefore has a shorter focal length and focuses images of near objects.

circadian rhythm

Metabolic rhythm found in most organisms, which generally coincides with the twenty-four-hour day. Its most obvious manifestation is the regular cycle of sleeping and waking, but body temperature and the concentration of hormones that influence mood and behavior also vary over the day. In humans, alteration of habits (such as rapid air travel around the world) may result in the circadian rhythm's being out of phase with actual activity patterns, causing malaise until it has had time to adjust.

circulatory system

System of vessels in an animal's body that transports essential substances (blood or other circulatory fluid) to and from the different parts of the body. It was first discovered and described by English physician William Harvey. All mammals except for the simplest kinds—such as sponges, jellyfish, sea anemones, and corals—have some type of circulatory system. Some invertebrates (animals without a backbone), such as insects, spiders, and most shellfish, have an "open" circulatory system, which consists of a simple network of tubes and hollow spaces. Other invertebrates have pumplike structures that send blood through a system of blood vessels. All vertebrates (animals with a backbone), including human beings, have a "closed" circulatory system that principally consists of a pumping organ—the heart—and a network of blood vessels.

cistron

In genetics, the segment of DNA that is required to synthesize a complete polypeptide chain. It is the molecular equivalent of a gene.

CITES

See **Convention on International Trade in Endangered Species.**

citric acid cycle

See **Krebs cycle.**

cladistics

Method of biological classification (taxonomy) that uses a formal step-by-step procedure for objectively assessing the extent to which organisms share

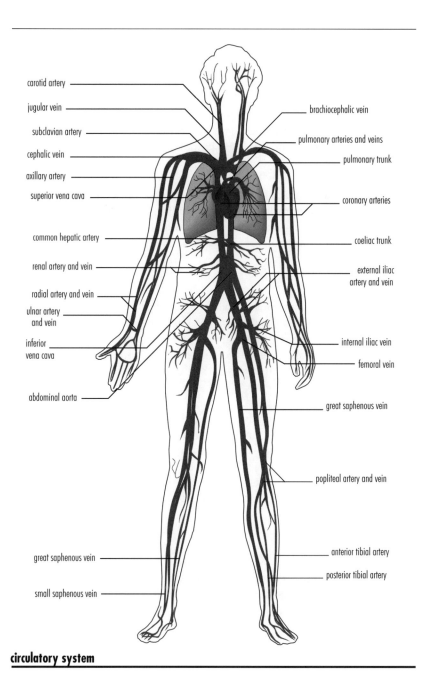

carotid artery

jugular vein

subclavian artery

cephalic vein

axillary artery

superior vena cava

common hepatic artery

renal artery and vein

radial artery and vein

ulnar artery and vein

inferior vena cava

abdominal aorta

brachiocephalic vein

pulmonary arteries and veins

pulmonary trunk

coronary arteries

coeliac trunk

external iliac artery and vein

internal iliac vein

femoral vein

great saphenous vein

popliteal artery and vein

anterior tibial artery

posterior tibial artery

great saphenous vein

small saphenous vein

circulatory system

particular characters, and for assigning them to taxonomic groups. Taxonomic groups (for example, species, genus, family) are termed clades.

cladode

In botany, a flattened stem that is leaflike in appearance and function. It is an adaptation to dry conditions because a stem contains fewer stomata than a leaf, and water loss is thus minimized. The true leaves in such plants are usually reduced to spines or small scales. Examples of plants with cladodes are butcher's-broom *Ruscus aculeatus,* asparagus, and certain cacti. Cladodes may bear flowers or fruit on their surface, and this distinguishes them from leaves.

class

In biological classification, a group of related orders. For example, all mammals belong to the class Mammalia and all birds to the class Aves. Among plants, all class names end in "idae" (such as Asteridae) and among fungi in "mycetes"; there are no equivalent conventions among animals. Related classes are grouped together in a phylum.

classification

The arrangement of organisms into a hierarchy of groups on the basis of their similarities in biochemical, anatomical, or physiological characters. The basic grouping is a species, several of which may constitute a genus, which in turn are grouped into families, and so on up through orders, classes, phyla (in plants, sometimes called divisions), to kingdoms.

clavicle

From Latin *clavis,* meaning "key," the collarbone of many vertebrates. In humans it is vulnerable to fracture, since falls involving a sudden force on the arm may result in very high stresses passing into the chest region by way of the clavicle and other bones. It is connected at one end with the sternum (breastbone), and at the other end with the shoulder blade, together with which it forms the arm socket. The wishbone of a chicken is composed of its two fused clavicles.

claw

Hard, hooked, pointed outgrowth of the digits of mammals, birds, and most reptiles. Claws are composed of the protein keratin and grow continuously from a bundle of cells in the lower skin layer. Hooves and nails are modified structures with the same origin as claws.

cleistogamy

Production of flowers that never fully open and that are automatically self-fertilized. Cleistogamous flowers are often formed late in the year, after the production of normal flowers, or during a period of cold weather, as seen in several species of violet *Viola.*

climax community

Assemblage of plants and animals that is relatively stable in its environment. It is brought about by ecological succession and represents the point at which succession ceases to occur.

cloaca

The common posterior chamber of most vertebrates into which the digestive, urinary, and reproductive tracts all enter; a cloaca is found in most reptiles, birds, and amphibians; many fishes; and, to a reduced degree, marsupial mammals. Placental mammals, however, have a separate digestive opening (the anus) and urinogenital opening. The cloaca forms a chamber in which products can be stored before being voided from the body via a muscular opening, the cloacal aperture.

clone

An exact replica. In genetics, any one of a group of genetically identical cells or organisms. An identical twin is a clone; so, too, are bacteria living in the same colony. The term *clone* has been adopted by computer technology to describe a (nonexistent) device that mimics an actual one to enable certain software programs to run correctly.

coccus

Plural *cocci*, member of a group of globular bacteria, some of which are harmful to humans. The cocci contain the subgroups streptococci, where the bacteria associate in straight chains, and staphylococci, where the bacteria associate in branched chains.

cochlea

Part of the inner ear. It is equipped with approximately ten thousand hair cells, which move in response to sound waves and thus stimulate nerve cells to send messages to the brain. In this way they turn vibrations of the air into electrical signals.

codominance

In genetics, the failure of a pair of alleles, controlling a particular characteristic, to show the classic recessive-dominant relationship. Instead, aspects of both alleles may show in the phenotype.

codon

In genetics, a triplet of bases in a molecule of DNA or RNA that directs the placement of a particular amino acid during the process of protein (polypeptide) synthesis. There are sixty-four codons in the genetic code.

coefficient of relationship

The probability that any two individuals share a given gene by virtue of being descended from a common ancestor. In sexual reproduction of diploid species, an individual shares half its genes with each parent, with its offspring, and (on average) with each sibling; but only a quarter (on average) with its grandchildren or its siblings' offspring; an eighth with its great-grandchildren, and so on.

coelenterate

Any freshwater or marine organism of the phylum Coelenterata, having a body wall composed of two layers of cells. They also possess stinging cells. Examples are jellyfish, hydra, and coral.

coelom

In all but the simplest animals, the fluid-filled cavity that separates the body wall from the gut and associated organs and allows the gut muscles to contract independently of the rest of the body.

coevolution

Evolution of those structures and behaviors within a species that can best be understood in relation to another species. For example, insects and flowering plants have evolved together: insects have produced mouthparts suitable for collecting pollen or drinking nectar, and plants have developed chemicals and flowers that will attract insects to them.

cold-blooded

A descriptor for animals whose temperature is not internally regulated but depends on the surrounding temperature; *see* **poikilothermy.**

coleoptile

The protective sheath that surrounds the young shoot tip of a grass during its passage through the soil to the surface. Although of relatively simple structure, most coleoptiles are very sensitive to light, ensuring that seedlings grow upward.

collagen

Protein that is the main constituent of connective tissue. Collagen is present in skin, cartilage, tendons, and ligaments. Bones are made up of collagen, with the mineral calcium phosphate providing increased rigidity.

collenchyma

Plant tissue composed of relatively elongated cells with thickened cell walls, in particular at the corners where adjacent cells meet. It is a supporting and strengthening tissue found in nonwoody plants, mainly in the stems and leaves.

colon

In anatomy, the main part of the large intestine, between the cecum and rectum. Water and mineral salts are absorbed from undigested food in the colon, and the residue passes as feces toward the rectum.

colonization

In ecology, the spread of species into a new habitat, such as a freshly cleared field, the shoulder of a new highway, or a recently flooded valley. The first species to move in are called pioneers and may establish conditions that allow other animals and plants to move in (for example, by improving the condition of the soil or by providing shade). Over time a range of species arrives and the habitat matures; early colonizers will probably be replaced, so that the variety of animal and plant life present changes. This is known as succession.

commensalism

A relationship between two species whereby one (the commensal) benefits from the association, whereas the other neither benefits nor suffers. For example, certain species of millipede and silverfish inhabit the nests of army ants and live by scavenging on the refuse of their hosts, but without affecting the ants.

communication

The signaling of information by one organism to another, usually with the intention of altering the recipient's behavior. Signals used in communication may be visual (such as the human smile or the display of colorful plumage in birds), auditory (for example, the whines or barks of a dog), olfactory (such as the odors released by the scent glands of a deer), electrical (as in the pulses emitted by electric fish), or tactile (for example, the nuzzling of male and female elephants).

community

In ecology, an assemblage of plants, animals, and other organisms living within a circumscribed area. Communities are usually named by reference to a dominant feature such as characteristic plant species (for example, a beech-wood community), or a prominent physical feature (for example, a freshwater-pond community).

comparative anatomy

Study of the similarity and differences in the anatomy of different groups of animals. Such study helps to reveal how animals are related to each other and how they have changed through evolution.

compensation point

The point at which there is just enough light for a plant to survive. At this point all the food produced by photosynthesis is used up by respiration. For aquatic plants, the compensation point is the depth of water at which there is just enough light to sustain life (deeper water = less light = less photosynthesis).

competition

In ecology, the interaction between two or more organisms, or groups of organisms (for example, species), that use a common resource that is in short supply.

Competition invariably results in a reduction in the numbers of one or both competitors, and in evolution contributes both to the decline of certain species and to the evolution of adaptations.

complementation
In genetics, the interaction that can occur between two different mutant alleles of a gene in a diploid organism to make up for each other's deficiencies and allow the organism to function normally.

Compositae
Daisy family, comprising dicotyledonous flowering plants characterized by flowers borne in composite heads. It is the largest family of flowering plants, the majority being herbaceous. Birds seem to favor the family for use in nest "decoration," possibly because many species either repel or kill insects. Species include the daisy and dandelion; food plants such as the artichoke, lettuce, and safflower; and the garden varieties of chrysanthemum, dahlia, and zinnia.

cone
In botany, the reproductive structure of the conifers and cycads; also known as a strobilus. It consists of a central axis surrounded by numerous, overlapping, scalelike, modified leaves (sporophylls) that bear the reproductive organs. Usually there are separate male and female cones, the former bearing pollen sacs containing pollen grains, and the larger female cones bearing the ovules that contain the ova or egg cells. The pollen is carried from male to female cones by the wind (anemophily). The seeds develop within the female cone and are released as the scales open in dry atmospheric conditions, which favor seed dispersal.

congenital disease
In medicine, a disease that is present at birth. It is not necessarily genetic in origin; for example, congenital herpes may be acquired by the baby as it passes through the mother's birth canal.

conjugation
The bacterial equivalent of sexual reproduction. A fragment of the DNA from one bacterium is passed along a thin tube, the pilus, into another bacterium.

conjunctiva
Membrane covering the front of the vertebrate eye. It is continuous with the epidermis of the eyelids and lies on the surface of the cornea.

connective tissue
In animals, tissue made up of a noncellular substance, the extracellular matrix, in which some cells are embedded. Skin, bones, tendons, cartilage, and adipose tissue (fat) are the main connective tissues. There are also small amounts of con-

nective tissue in organs such as the brain and liver, where they maintain shape and structure.

conservation

In the life sciences, action taken to protect and preserve the natural world, usually from pollution, overexploitation, and other harmful features of human activity. The late 1980s saw a great increase in public concern for the environment, with membership in conservation groups, such as Friends of the Earth, Greenpeace, and the Sierra Club rising sharply. Globally the most important issues include the depletion of atmospheric ozone by the action of chlorofluorocarbons (CFCs), the buildup of carbon dioxide in the atmosphere (thought to contribute to an intensification of the greenhouse effect), and deforestation.

contraceptive

Any drug, device, or technique that prevents pregnancy. The contraceptive pill ("the Pill") contains female hormones that interfere with egg production or the first stage of pregnancy. The "morning-after" pill can be taken up to seventy-two hours after unprotected intercourse. Barrier contraceptives include condoms (sheaths) and diaphragms, also called caps or Dutch caps; they prevent the sperm from entering the cervix (neck of the womb). Intrauterine devices, also known as IUDs or coils, cause a slight inflammation of the lining of the womb; this prevents the fertilized egg from becoming implanted.

contractile root

In botany, a thickened root at the base of a corm, bulb, or other organ that helps position it at an appropriate level in the ground. Contractile roots are found, for example, on the corms of plants of the genus *Crocus*. After they have become anchored in the soil, the upper portion contracts, pulling the plant deeper into the ground.

control experiment

Essential part of a scientifically valid experiment, designed to show that the factor being tested is actually responsible for the effect observed. In the control experiment all factors, apart from the one under test, are exactly the same as in the test experiments, and all the same measurements are carried out. For example, in drug trials, a placebo (a harmless substance) is given alongside the substance being tested in order to compare effects.

Convention on International Trade in Endangered Species (CITES)

International agreement under the auspices of the International Union for the Conservation of Nature (IUCN) regulating trade in endangered species of animals and plants. The agreement came into force in 1975 and by 1997 had been signed by 138 states. It prohibits any trade in a category of eight thousand highly endangered species and controls trade in a further thirty thousand species.

Animals and plants listed in Appendix 1 of CITES are classified endangered; those listed in Appendix 2 are classified vulnerable.

convergent evolution
The independent evolution of similar structures in species (or other taxonomic groups) that are not closely related, as a result of living in a similar way. Thus, birds and bats have wings, not because they are descended from a common winged ancestor, but because their respective ancestors independently evolved flight.

copepod
Crustacean of the subclass Copepoda, mainly microscopic and found in plankton.

copulation
Act of mating in animals with internal fertilization. Male mammals have a penis or other organ that is used to introduce spermatozoa into the reproductive tract of the female. Most birds transfer sperm by pressing their cloacas (the openings of their reproductive tracts) together.

cork
Light, waterproof outer layers of the bark covering the branches and roots of almost all trees and shrubs. The cork oak *(Quercus suber),* a native of southern Europe and northern Africa, is cultivated in Spain and Portugal; the exceptionally thick outer layers of its bark provide the cork that is used commercially.

corm
Short, swollen, underground plant stem, surrounded by protective scale leaves, as seen in the genus *Crocus.* It stores food, provides a means of vegetative reproduction, and acts as a perennating organ.

cornea
Transparent front section of the vertebrate eye. The cornea is curved and behaves as a fixed lens, so that light entering the eye is partly focused before it reaches the lens.

corolla
Collective name for the petals of a flower. In some plants the petal margins are partly or completely fused to form a corolla tube, for example in bindweed *Convolvulus arvensis.*

corpus luteum
Glandular tissue formed in the mammalian ovary after ovulation from the Graafian follicle, a group of cells associated with bringing the egg to maturity. It secretes the hormone progesterone in anticipation of pregnancy.

cortex

The outer part of a structure such as the brain, kidney, or adrenal gland. In botany the cortex includes nonspecialized cells lying just beneath the surface cells of the root and stem.

cotyledon

Structure in the embryo of a seed plant that may form a "leaf" after germination and is commonly known as a seed leaf. The number of cotyledons present in an embryo is an important factor in the classification of flowering plants (angiosperms).

courtship

Behavior exhibited by animals as a prelude to mating. The behavior patterns vary considerably from one species to another, but are often ritualized forms of behavior not obviously related to courtship or mating (for example, courtship feeding in birds).

cranium

The dome-shaped area of the vertebrate skull that protects the brain. It consists of eight bony plates fused together by sutures (immovable joints). Fossil remains of the human cranium have aided the development of theories concerning human evolution.

creationism

Theory concerned with the origins of matter and life, claiming, as does the Bible in Genesis, that the world and humanity were created by a supernatural Creator, not more than six thousand years ago. The theory was developed in response to Darwin's theory of evolution; it is not recognized by most scientists as having a factual basis.

crop

In birds, the thin-walled enlargement of the digestive tract between the esophagus and stomach. It is an effective storage organ especially in seed-eating birds; a pigeon's crop can hold about five hundred cereal grains. Digestion begins in the crop, by the moisturizing of food. A crop also occurs in insects and annelid worms.

crustacean

One of the class of arthropods that includes crabs, lobsters, shrimps, wood lice, and barnacles. The external skeleton is made of protein and chitin hardened with lime. Each segment bears a pair of appendages that may be modified as sensory feelers (antennae), as mouthparts, or as swimming, walking, or grasping structures.

cryptogam

Obsolete name applied to the lower plants. It included the algae, liverworts, mosses, and ferns (plus the fungi and bacteria in very early schemes of classification). In such classifications seed plants were known as phanerogams.

cryptosporidium

Waterborne parasite that causes disease in humans and other animals. It has been found in drinking water in the United States; it causes diarrhea, abdominal cramps, vomiting, and fever and can be fatal in people with damaged immune systems, such as AIDS sufferers or those with leukemia. As few as thirty cryptosporidia are enough to cause prolonged diarrhea.

cultivar

Variety of a plant developed by horticultural or agricultural techniques. The term derives from "*culti*vated *var*iety."

culture

The growing of living cells and tissues in laboratory conditions.

cuticle

The horny noncellular surface layer of many invertebrates such as insects; in botany, the waxy surface layer on those parts of plants that are exposed to the air, continuous except for stomata and lenticels. All types are secreted by the cells of the epidermis. A cuticle reduces water loss and, in arthropods, acts as an exoskeleton.

cutting

Technique of vegetative propagation involving taking a section of root, stem, or leaf and treating it so that it develops into a new plant.

cyanobacteria

(Singular *cyanobacterium*), *see* **blue-green algae.**

cyanocobalamin

Chemical name for vitamin B_{12}, which is normally produced by microorganisms in the gut. The richest sources are liver, fish, and eggs. It is essential to the replacement of cells, the maintenance of the myelin sheath, which insulates nerve fibers, and the efficient use of folic acid, another vitamin in the B complex. Deficiency can result in pernicious anemia (defective production of red blood cells) and possible degeneration of the nervous system.

cytochrome

Protein responsible for part of the process of respiration by which food molecules are broken down in aerobic organisms. Cytochromes are part of the electron transportation chain, which uses energized electrons to reduce molecular oxygen (O_2) to oxygen ions (O^{2-}). These combine with hydrogen ions (H^+) to form water (H_2O), the end product of aerobic respiration. As electrons are passed from one cytochrome to another, energy is released and used to make adenosine triphosphate (ATP).

cytokinin

Plant hormone that stimulates cell division. Cytokinins affect several different aspects of plant growth and development, but only if auxin is also present. They may delay the process of senescence, or aging, break the dormancy of certain seeds and buds, and induce flowering.

cytology

The study of cells and their functions. Major advances have been made possible in this field by the development of the electron microscope.

cytoplasm

The part of the cell outside the nucleus. Strictly speaking, this includes all the organelles (mitochondria, chloroplasts, and so on), but often cytoplasm refers to the jellylike matter in which the organelles are embedded (correctly termed the cytosol). The cytoplasm is the site of protein synthesis.

cytoskeleton

In a living cell, a matrix of protein filaments and tubules that occurs within the cytosol (the liquid part of the cytoplasm). It gives the cell a definite shape, transports vital substances around the cell, and may also be involved in cell movement.

DDT

Abbreviation for dichloro-diphenyl-trichloroethane ($ClC_6H_5)_2CHC(HCl_2$), insecticide discovered in 1939 by Swiss chemist Paul Müller. It is useful in the control of insects that spread malaria, but resistant strains develop. DDT is highly toxic and persists in the environment and in living tissue. Its use is now banned in most countries, but it continues to be used on food plants in Latin America.

deamination

Removal of the amino group ($-NH_2$) from an unwanted amino acid. This is the nitrogen-containing part, and it is converted into ammonia, uric acid, or urea (depending on the type of animal) to be excreted in the urine. In vertebrates, deamination occurs in the liver.

death

Cessation of all life functions, so that the molecules and structures associated with living things become disorganized and indistinguishable from similar molecules found in nonliving things. In medicine, a person is pronounced dead when the brain ceases to control the vital functions, even if breathing and heartbeat are maintained artificially.

deciduous

Of trees and shrubs, designating those that shed their leaves at the end of the growing season or during a dry season to reduce transpiration (the loss of water by evaporation).

deciduous teeth

See **milk teeth.**

decomposer

Any organism that breaks down dead matter. Decomposers play a vital role in the ecosystem by freeing important chemical substances, such as nitrogen compounds, locked up in dead organisms or excrement. They feed on some of the released organic matter, but leave the rest to filter back into the soil as dissolved nutrients, or pass in gas form into the atmosphere, for example as nitrogen and carbon dioxide. The principal decomposers are bacteria and fungi, but earthworms and many other invertebrates are often included in this group. The nitrogen cycle relies on the actions of decomposers.

decomposition

Process whereby a chemical compound is reduced to its component substances. Decomposition results in the destruction of dead organisms either by chemical reduction or by the action of decomposers, such as bacteria and fungi.

degeneration

A change in the structure or chemical composition of a tissue or organ that interferes with its normal functioning. Examples of degeneration include fatty degeneration, fibroid degeneration (cirrhosis), and calcareous degeneration, all of which are part of natural changes that occur in old age.

dendrite

Part of a nerve cell or neuron. The dendrites are slender filaments projecting from the cell body. They receive incoming messages from many other nerve cells and pass them on to the cell body. If the combined effect of these messages is strong enough, the cell body will send an electrical impulse along the axon (the threadlike extension of a nerve cell). The tip of the axon passes its message to the dendrites of other nerve cells.

dendrochronology, or tree-ring dating

Analysis of the annual rings of trees to date past events by determining the age of timber. Since annual rings are formed by variations in the water-conducting cells produced by the plant during different seasons of the year, they also provide a means of establishing past climatic conditions in a given area.

denitrification
Process occurring naturally in soil, where bacteria break down nitrates to give nitrogen gas, which returns to the atmosphere.

dental formula
Way of showing the number of teeth in an animal's mouth. The dental formula consists of eight numbers separated by a horizontal line into two rows. The four above the line represent the teeth on one side of the upper jaw, starting at the front. If this reads 2 1 2 3 (as for humans) it means two incisors, one canine, two premolars, and three molars. The numbers below the line represent the lower jaw. The total number of teeth can be calculated by adding up all the numbers and multiplying by two.

dentition
Type and number of teeth in a species. Different kinds of teeth have different functions; a grass-eating animal will have large molars for grinding its food, whereas a meat-eater will need powerful canines for catching and killing its prey. The teeth that are less useful to an animal's lifestyle may be reduced in size or missing altogether. An animal's dentition is represented diagrammatically by a dental formula.

deoxyribonucleic acid
See **DNA.**

detritus
The organic debris produced during the decomposition of animals and plants.

development
The process whereby a living thing transforms itself from a single cell into a vastly complicated multicellular organism, with structures, such as limbs, and functions, such as respiration, all able to work correctly in relation to each other. Most of the details of this process remain unknown, although some of the central features are becoming understood.

dextrose
See **glucose.**

diabetes mellitus
Disease in which a disorder of the islets of Langerhans in the pancreas prevents the body producing the hormone insulin, so that sugars cannot be used properly. Treatment is by strict dietary control and oral or injected insulin, depending on the type of diabetes.

diapause

Period of suspended development that occurs in some species of insects, characterized by greatly reduced metabolism. Periods of diapause are often timed to coincide with the winter months and improve the insect's chances of surviving adverse conditions.

diaphragm

In mammals, a thin muscular sheet separating the thorax from the abdomen. It is attached by way of the ribs at either side and the breastbone and backbone, and a central tendon. Arching upward against the heart and lungs, the diaphragm is important in the mechanics of breathing. It contracts at each inhalation, moving downward to increase the volume of the chest cavity, and relaxes at exhalation.

diastole

The relaxation of a hollow organ. In particular, the term is used to indicate the resting period between beats of the heart when blood is flowing into it.

diastolic pressure

In medicine, measurement due to the pressure of blood against the arterial wall during diastole (relaxation of the heart). It is the lowest blood pressure during the cardiac cycle. The average diastolic pressure in healthy young adults is about 80 mmHg. The variation of diastolic pressure due to changes in body position and mood is greater than that of systolic pressure. Diastolic pressure is also a more accurate predictor of hypertension (high blood pressure).

dichloro-diphenyl-trichloroethane

See **DDT.**

dicotyledon

Major subdivision of the angiosperms, containing the great majority of flowering plants. Dicotyledons are characterized by the presence of two seed leaves, or cotyledons, in the embryo, which is usually surrounded by the endosperm. They generally have broad leaves with netlike veins.

diet

Range of foods eaten by an animal each day; it is also a particular selection of food, or the total amount and choice of food for a specific person or people. Most animals require seven kinds of food in their diet: proteins, carbohydrates, fats, vitamins, minerals, water, and roughage. A diet that contains all of these things in the correct amounts and proportions is termed a balanced diet. The amounts and proportions required varies with different animals, according to their size, age, and lifestyle. The digestive systems of animals have evolved to meet particular needs; they have also adapted to cope with the foods available in the surroundings in which they live. The necessity of finding and processing an appropriate diet is a very basic drive in animal evolution. Dietetics is the science of feeding individuals or groups; a dietition is a specialist in this science.

differentiation

In embryology, the process by which cells become increasingly different and specialized, giving rise to more complex structures that have particular functions in the adult organism. For instance, embryonic cells may develop into nerve, muscle, or bone cells.

diffusion

Spontaneous and random movement of molecules or particles in a fluid (gas or liquid) from a region in which they are at a high concentration to a region of lower concentration, until a uniform concentration is achieved throughout. The difference in concentration between two such regions is called the concentration gradient. No mechanical mixing or stirring is involved. For instance, if a drop of ink is added to water, its molecules will diffuse until their color becomes evenly distributed throughout. Diffusion occurs more rapidly across a higher concentration gradient and at higher temperatures.

digestion

Process whereby food eaten by an animal is broken down mechanically, and chemically by enzymes, mostly in the stomach and intestines, to make the nutrients available for absorption and cell metabolism.

digestive system

In the body, all the organs and tissues involved in the digestion of food. In animals, these consist of the mouth, stomach, intestines, and their associated glands. The process of digestion breaks down the food by physical and chemical means into the different elements that are needed by the body for energy and tissue building and repair. Digestion begins in the mouth and is completed in the stomach; from there most nutrients are absorbed into the small intestine from where they pass through the intestinal wall into the bloodstream; what remains is stored and concentrated into feces in the large intestine. Birds have two additional digestive organs—the crop and gizzard. In smaller, simpler animals such as jellyfish, the digestive system is simply a cavity (coelenteron or enteric cavity) with a "mouth" into which food is taken; the digestible portion is dissolved and absorbed in this cavity, and the remains are ejected back through the mouth.

dihybrid inheritance

In genetics, a pattern of inheritance observed when two characteristics are studied in succeeding generations. The first experiments of this type, as well as in monohybrid inheritance, were carried out by Austrian biologist Gregor Mendel using pea plants.

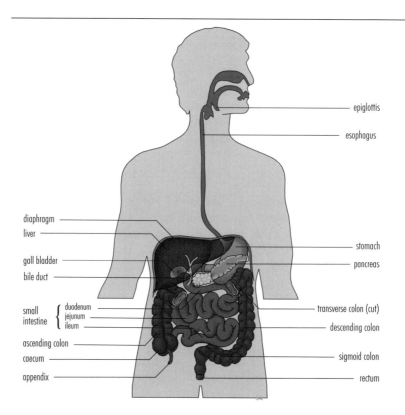

digestive system

dioecious

A descriptor of plants with male and female flowers borne on separate individuals of the same species. Dioecism occurs, for example, in the willows *Salix*. It is a way of avoiding self-fertilization.

diploblastic

Having a body wall composed of two layers. The outer layer is the ectoderm, the inner layer is the endoderm. This pattern of development is shown by coelenterates.

diploid

Having paired chromosomes in each cell. In sexually reproducing species, one set is derived from each parent, the gametes, or sex cells, of each parent being haploid (having only one set of chromosomes) due to meiosis (reduction cell division).

disaccharide

Sugar made up of two monosaccharides or simple sugars. Sucrose ($C_{12}H_{22}O_{11}$), or table sugar, is a disaccharide.

disease

Condition that disturbs or impairs the normal state of an organism. Diseases can occur in all life forms and normally affect the functioning of cells, tissues, organs, or systems. Diseases are usually characterized by specific symptoms and signs and can be mild and short-lasting—such as the common cold—or severe enough to decimate a whole species—such as Dutch elm disease. Diseases can be classified as infectious or noninfectious. Infectious diseases are caused by microorganisms, such as bacteria and viruses, invading the body; they can be spread across a species, or transmitted between one or more species. All other diseases can be grouped together as noninfectious diseases. These can have many causes: they may be inherited (congenital diseases); they may be caused by the ingestion or absorption of harmful substances, such as toxins; they can result from poor nutrition or hygiene; or they may arise from injury or aging. The causes of some diseases are still unknown.

dispersal

Phase of reproduction during which gametes, eggs, seeds, or offspring move away from the parents into other areas. The result is that overcrowding is avoided and parents do not find themselves in competition with their own offspring. The mechanisms are various, including a reliance on wind or water currents and, in the case of animals, locomotion. The ability of a species to spread widely through an area and to colonize new habitats has survival value in evolution.

displacement activity

In animal behavior, an action that is performed out of its normal context, while the animal is in a state of stress, frustration, or uncertainty. Birds, for example, often peck at grass when uncertain whether to attack or flee from an opponent; similarly, humans scratch their heads when nervous.

DNA

Abbreviation for deoxyribonucleic acid, complex giant molecule that contains, in chemically coded form, the information needed for a cell to make proteins. DNA is a ladderlike double-stranded nucleic acid that forms the basis of genetic inheritance in all organisms, except for a few viruses that have only RNA. DNA is organized into chromosomes and, in organisms other than bacteria, it is found only in the cell nucleus.

DNA fingerprinting, or DNA profiling

See **genetic fingerprinting.**

dominance

In genetics, the masking of one allele (an alternative form of a gene) by another allele. For example, if a heterozygous person has one allele for blue eyes and one for brown eyes, his or her eye color will be brown. The allele for blue eyes is described as recessive and the allele for brown eyes as dominant.

dopamine

Neurotransmitter, hydroxytyramine $C_8H_{11}NO_2$, an intermediate in the formation of adrenaline. There are special nerve cells (neurons) in the brain that use dopamine for the transmission of nervous impulses. One such area of dopamine neurons lies in the basal ganglia, a region that controls movement. Patients suffering from the tremors of Parkinson's disease show nerve degeneration in this region. Another dopamine area lies in the limbic system, a region closely involved with emotional responses. It has been found that schizophrenic patients respond well to drugs that limit dopamine excess in this area.

dormancy

In botany, a phase of reduced physiological activity exhibited by certain buds, seeds, and spores. Dormancy can help a plant to survive unfavorable conditions, as in annual plants that pass the cold winter season as dormant seeds, and plants that form dormant buds.

dorsal

In vertebrates, the surface of the animal closest to the backbone. For most vertebrates and invertebrates this is the upper surface, or the surface furthest from the ground. For bipedal primates such as humans, where the dorsal surface faces backward, then the word is "back."

drupe

Fleshy fruit containing one or more seeds that are surrounded by a hard, protective layer—for example cherry, almond, and plum. The wall of the fruit (pericarp) is differentiated into the outer skin (exocarp), the fleshy layer of tissues (mesocarp), and the hard layer surrounding the seed (endocarp).

ductless gland

Alternative name for an endocrine gland.

duodenum

In vertebrates, a short length of alimentary canal found between the stomach and the small intestine. Its role is in digesting carbohydrates, fats, and proteins. The smaller molecules formed are then absorbed, either by the duodenum or the ileum.

ear

Organ of hearing in animals. It responds to the vibrations that constitute sound, which are translated into nerve signals and passed to the brain. A mammal's ear consists of three parts: outer ear, middle ear, and inner ear. The outer ear is a funnel that collects sound, directing it down a tube to the eardrum (tympanic membrane), which separates the outer and middle ears. Sounds vibrate this membrane, the mechanical movement of which is transferred to a smaller membrane leading to the inner ear by three small bones, the auditory ossicles. Vibrations of the inner ear membrane move fluid contained in the snail-shaped cochlea, which vibrates hair cells that stimulate the auditory nerve connected to the brain. There are approximately thirty thousand sensory hair cells (stereocilia). Exposure to loud noise and the process of aging damages the stereocilia, resulting in hearing loss. Three fluid-filled canals of the inner ear detect changes of position; this mechanism, with other sensory inputs, is responsible for the sense of balance.

ecdysis
Periodic shedding of the exoskeleton by insects and other arthropods to allow growth. Prior to shedding, a new soft and expandable layer is laid down underneath the existing one. The old layer then splits, the animal moves free of it, and the new layer expands and hardens.

echinoderm
Marine invertebrate of the phylum Echinodermata ("spiny-skinned"), characterized by a five-radial symmetry. Echinoderms have a water-vascular system that transports substances around the body. They include starfishes (or sea stars), brittle-stars, sea lilies, sea urchins, and sea cucumbers. The skeleton is external, made of a series of limy plates. Echinoderms generally move by using tube-feet, small water-filled sacs that can be protruded or pulled back to the body.

echolocation, or biosonar
Method used by certain animals, notably bats, whales, and dolphins, to detect the positions of objects by using sound. The animal emits a stream of high-pitched sounds, generally at ultrasonic frequencies (beyond the range of human hearing), and listens for the returning echoes reflected off objects to determine their exact location.

E. coli
See *Escherichia coli.*

ecology
From Greek *oikos,* meaning "house," the study of the relationship among organisms and the environments in which they live, including all living and nonliving components. The chief environmental factors governing the distribution of plants and animals are temperature, humidity, soil, light intensity, day length, food supply, and interaction with other organisms. The term was coined by the biologist Ernst Haeckel in 1866.

ecosystem
In ecology, an integrated unit consisting of a community of living organisms—bacteria, animals, and plants—and the physical environment—air, soil, water, and climate—that they inhabit. Individual organisms interact with each other and with their environment, or habitat, in a series of relationships that depends on the flow of energy and nutrients through the system. These relationships are usually complex and finely balanced, and in theory natural ecosystems are self-sustaining. However, major changes to an ecosystem, such as climate change, overpopulation, or the removal of a species, may threaten the system's sustainability and result in its eventual destruction. For instance, the removal of a major carnivore predator can result in the destruction of an ecosystem through overgrazing by herbivores.

ectoparasite
Parasite that lives on the outer surface of its host.

ectoplasm
Outer layer of a cell's cytoplasm.

ectotherm
A "cold-blooded" animal, such as a lizard, that relies on external warmth (ultimately from the sun) to raise its body temperature so that it can become active. To cool the body, ectotherms seek out a cooler environment.

egg
In animals, the ovum, or female gamete (reproductive cell). After fertilization by a sperm cell, it begins to divide to form an embryo. Eggs may be deposited by the female (ovipary) or they may develop within her body (vivipary and ovovivipary). In the oviparous reptiles and birds, the egg is protected by a shell, and well supplied with nutrients in the form of yolk.

elytra
Horny wing cases characteristic of beetles. The elytra are adapted from the beetles' forewings (only the hindwings are used for flight). They fold over the back, generally meeting in the middle in a straight line, and serve to protect the hindwings and the soft posterior parts of the body.

embryo
Early developmental stage of an animal or a plant following fertilization of an ovum (egg cell), or activation of an ovum by parthenogenesis. In humans, the term embryo describes the fertilized egg during its first seven weeks of existence; from the eighth week onward it is referred to as a fetus.

embryology
Study of the changes undergone by an organism from its conception as a fertilized ovum (egg) to its emergence into the world at hatching or birth. It is mainly con-

cerned with the changes in cell organization in the embryo and the way in which these lead to the structures and organs of the adult (the process of differentiation).

embryo sac
Large cell within the ovule of flowering plants that represents the female gameto-phyte when fully developed. It typically contains eight nuclei. Fertilization occurs when one of these nuclei, the egg nucleus, fuses with a male gamete.

emulsifier
Food additive used to keep oils dispersed and in suspension in products such as mayonnaise and peanut butter. Egg yolk is a naturally occurring emulsifier, but most of the emulsifiers in commercial use today are synthetic chemicals.

encephalin
A naturally occurring chemical produced by nerve cells in the brain that has the same effect as morphine or other derivatives of opium, acting as a natural pain-killer. Unlike morphine, encephalins are quickly degraded by the body, so there is no buildup of tolerance to them, and hence no addiction. Encephalins are a variety of peptides, as are endorphins, which have similar effects.

endangered species
Plant or animal species whose numbers are so few that it is at risk of becoming extinct. Officially designated endangered species are listed by the International Union for the Conservation of Nature (IUCN).

endocrine gland
Gland that secretes hormones into the bloodstream to regulate body processes. Endocrine glands are most highly developed in vertebrates, but are also found in other animals, notably insects. In humans the main endocrine glands are the pi-tuitary, thyroid, parathyroid, adrenal, pancreas, ovary, and testis.

endolymph
Fluid found in the inner ear, filling the central passage of the cochlea as well as the semicircular canals.

endoparasite
Parasite that lives inside the body of its host.

endoplasm
Inner, liquid part of a cell's cytoplasm.

endoplasmic reticulum (ER)
A membranous system of tubes, channels, and flattened sacs that forms com-partments within eukaryotic cells. It stores and transports proteins within cells and also carries various enzymes needed for the synthesis of fats. The

ribosomes, or the organelles that carry out protein synthesis, are attached to parts of the ER.

endorphin

Natural substance (a polypeptide) that modifies the action of nerve cells. Endorphins are produced by the pituitary gland and hypothalamus of vertebrates. They lower the perception of pain by reducing the transmission of signals between nerve cells.

endoskeleton

The internal supporting structure of vertebrates, made up of cartilage or bone. It provides support and acts as a system of levers to which muscles are attached to provide movement. Certain parts of the skeleton (the skull and ribs) give protection to vital body organs.

endosperm

Nutritive tissue in the seeds of most flowering plants. It surrounds the embryo and is produced by an unusual process that parallels the fertilization of the ovum by a male gamete. A second male gamete from the pollen grain fuses with two female nuclei within the embryo sac. Thus endosperm cells are triploid (having three sets of chromosomes); they contain food reserves such as starch, fat, and protein that are utilized by the developing seedling.

endotherm

A "warm-blooded," or homeothermic, animal. Endotherms have internal mechanisms for regulating their body temperatures to levels different from the environmental temperature.

endotoxin

Heat stable complex of protein and lipopolysaccharide that is produced following the death of certain bacteria. Endotoxins are typically produced by the Gram negative bacteria and can cause fever. They can also cause shock by rendering the walls of the blood vessels permeable so that fluid leaks into the tissues and blood pressure falls sharply.

entomology

The study of insects.

environment–heredity controversy

See **nature–nurture controversy.**

enzyme

Biological catalyst produced in cells and capable of speeding up the chemical reactions necessary for life. They are large, complex proteins and are highly specific, each chemical reaction requiring its own particular enzyme. The enzyme's specificity arises from its active site, an area with a shape corresponding to the part of the molecule with which it reacts (the substrate). The enzyme and the substrate slot together forming an enzyme-substrate complex that allows the reaction to take place, after which the enzyme falls away unaltered.

ephemeral plant
Plant with a very short life cycle, sometimes as little as six to eight weeks. It may complete several generations in one growing season. Many desert plants are ephemeral.

epidermis
Outermost layer of cells on an organism's body. In plants and many invertebrates such as insects, it consists of a single layer of cells. In vertebrates, it consists of several layers of cells.

epigeal
Seed germination in which the cotyledons (seed leaves) are borne above the soil.

epiglottis
Small flap located behind the root of the tongue in mammals. It closes off the end of the windpipe during swallowing to prevent food from passing into it and causing choking.

epinephrine
See **adrenalin.**

epiphyte
Any plant that grows on another plant or object above the surface of the ground and has no roots in the soil. An epiphyte does not parasitize the plant it grows on but merely uses it for support. Its nutrients are obtained from rainwater, organic debris such as leaf litter, or from the air.

epithelium
In animals, tissue of closely packed cells that forms a surface or lines a cavity or tube. The epithelium may be protective (as in the skin) or secretory (as in the cells lining the wall of the gut).

erythrocyte
See **red blood cell.**

Escherichia coli (E. coli)
Rod-shaped Gram-negative bacterium that lives, usually harmlessly, in the colon of most warm-blooded animals. It is the commonest cause of urinary tract infections in humans. It is sometimes found in water or meat where fecal

contamination has occurred and can cause severe gastric problems. The mapping of the genome of *E. coli*, consisting of 4,403 genes, was completed in 1997.

esophagus

Muscular tube by which food travels from mouth to stomach. The human esophagus is about 23 cm/9 in long. It extends downward from the pharynx, immediately behind the windpipe. It is lined with a mucous membrane, which secretes lubricant fluid to assist the downward movement of food (peristalsis).

essential amino acid

Water-soluble organic molecule vital to a healthy diet.

essential fatty acid

Organic compound consisting of a hydrocarbon chain and important in the diet.

estivation

In zoology, a state of inactivity and reduced metabolic activity, similar to hibernation, that occurs during the dry season in species such as lungfish and snails. In botany, the term is used to describe the way in which flower petals and sepals are folded in the buds. It is an important feature in plant classification.

estradiol

A type of estrogen (female sex hormone) and the principal one in mammals.

estrogen

Any of a group of hormones produced by the ovaries of vertebrates; the term is also used for various synthetic hormones that mimic their effects. The principal estrogen in mammals is estradiol. Estrogens control female sexual development, promote the growth of female secondary sexual characteristics, stimulate egg production, and, in mammals, prepare the lining of the uterus for pregnancy.

estrus

In mammals, the period during a female's reproductive cycle (also known as the estrus cycle or menstrual cycle) when mating is most likely to occur. It usually coincides with ovulation.

ethology

Comparative study of animal behavior in its natural setting. Ethology is concerned with the causal mechanisms (both the stimuli that elicit behavior and the physiological mechanisms controlling it), as well as the development of behavior, its function, and its evolutionary history.

etiolation

In botany, a form of growth seen in plants receiving insufficient light. It is characterized by long, weak stems, small leaves, and a pale yellowish color (chlorosis) owing to a lack of chlorophyll. The rapid increase in height enables a plant that is surrounded by others to reach quickly a source of light, after which a return to normal growth usually occurs.

eubacteria

An order of bacteria with rigid cell walls but lacking photosynthetic pigments. There are thirteen families within the order.

eukaryote

One of the two major groupings into which all organisms are divided. Included are all organisms, except bacteria and cyanobacteria (blue-green algae), which belong to the prokaryote grouping.

eusociality

Form of social life found in insects such as honey bees and termites, in which the colony is made up of special castes (for example, workers, drones, and reproductives) whose membership is biologically determined. The worker castes do not usually reproduce. Only one mammal, the naked mole rat, has a social organization of this type. A eusocial shrimp was discovered in 1996 living in the coral reefs of Belize. *Synalpheus regalis* lives in colonies of up to three hundred individuals, all the offspring of a single reproductive female.

Eustachian tube

Small air-filled canal connecting the middle ear with the back of the throat. Found in all land vertebrates, it equalizes the pressure on both sides of the eardrum.

evergreen

In botany, a plant such as pine, spruce, or holly, that bears its leaves all year round. Most conifers are evergreen. Plants that shed their leaves in autumn or during a dry season are described as *deciduous*.

evolution

The slow, gradual process of change from one form to another, as in the evolution of the universe from its formation to its present state, or in the evolution of life on Earth. Evolution is the process by which life has developed by stages from single-celled organisms into the multiplicity of animal and plant life, extinct and existing, that inhabits the earth. The development of the concept of evolution is usually associated with the English naturalist Charles Darwin, who attributed the main role in evolutionary change to natural selection acting on randomly occurring variations. However, these variations in species are now known to be adaptations produced by spontaneous changes or mutations in the genetic material of organisms.

evolutionary stable strategy (ESS)

In sociobiology, an assemblage of behavioral or physical characters (collectively termed a *strategy*) of a population that is resistant to replacement by any forms bearing new traits, because the new traits will not be capable of successful reproduction.

excretion

The removal of the waste products of metabolism from living organisms. In plants and simple animals, waste products are removed by diffusion. Plants, for example, excrete O_2, a product of photosynthesis. In mammals, waste products are removed by specialized excretory organs, principally the kidneys, which excrete urea. Water and metabolic wastes are also excreted in the feces and, in humans, through the sweat glands in the skin; carbon dioxide and water are removed via the lungs. The liver excretes bile pigments.

exfoliation

The separation of pieces of dead bone or skin in layers.

exocrine gland

Gland that discharges secretions, usually through a tube or a duct, on to a surface. Examples include sweat glands, which release sweat on to the skin, and digestive glands, which release digestive juices on to the walls of the intestine. Some animals also have endocrine glands (ductless glands) that release hormones directly into the bloodstream.

exoskeleton

The hardened external skeleton of insects, spiders, crabs, and other arthropods. It provides attachment for muscles and protection for the internal organs, as well as support. To permit growth it is periodically shed in a process called ecdysis.

extensor

A muscle that straightens a limb.

extinction

The complete disappearance of a species or higher taxon. Extinctions occur when an animal becomes unfit for survival in its natural habitat, usually to be replaced by another, better-suited animal. An organism becomes ill-suited for survival because its environment is changed or because its relationship to other organisms is altered. For example, a predator's fitness for survival depends upon the availability of its prey.

extracellular matrix

Strong material naturally occurring in animals and plants, made up of protein and long-chain sugars (polysaccharides) in which cells are embedded. It is often called a "biological glue," and forms part of connective tissues such as bone and skin.

eye

The organ of vision. In the human eye, the light is focused by the combined action of the curved cornea, the internal fluids, and the lens. The insect eye is compound—made up of many separate facets—known as ommatidia, each of which collects light and directs it separately to a receptor to build up an image. Invertebrates have much simpler eyes, with no lenses. Among mollusks, cephalopods have complex eyes similar to those of vertebrates. The mantis shrimp's eyes contain ten color pigments with which to perceive color; some flies and fishes have five, while the human eye has only three.

Fallopian tube, or oviduct

In mammals, one of two tubes that carry eggs from the ovary to the uterus. An egg is fertilized by sperm in the Fallopian tubes, which are lined with cells whose cilia move the egg toward the uterus.

family

In biological classification, a group of related genera. Family names are not printed in italic (unlike genus and species names), and by convention they all have the ending -idae (animals) or -aceae (plants and fungi). For example, the genera of hummingbirds are grouped in the hummingbird family, Trochilidae. Related families are grouped together in an order.

fat

In the broadest sense, a mixture of lipids—chiefly triglycerides (lipids containing three fatty acid molecules linked to a molecule of glycerol). More specifically, the term refers to a lipid mixture that is solid at room temperature (20°C); lipid mixtures that are liquid at room temperature are called oils. The higher the proportion of saturated fatty acids in a mixture, the harder the fat.

fatigue

In muscle, reduced response brought about by the accumulation of lactic acid in muscle tissue due to excessive cellular activity.

fatty acid, or carboxylic acid

Organic compound consisting of a hydrocarbon chain, up to twenty-four carbon atoms long, with a carboxyl group (–COOH) at one end. The covalent bonds between the carbon atoms may be single or double; where a double bond occurs the carbon atoms concerned carry one instead of two hydrogen atoms. Chains with only single bonds have all the hydrogen they can carry, so they are said to be saturated with hydrogen. Chains with one or more double bonds are said to be unsaturated. Fatty acids are produced in the small intestine when fat is digested.

feather

Rigid outgrowth of the outer layer of the skin of birds, made of the protein keratin. Feathers provide insulation and facilitate flight. There are several types, including long quill feathers on the wings and tail, fluffy down feathers for retaining body heat, and contour feathers covering the body. The coloring of feathers is often important in camouflage or in courtship and other displays. Feathers are normally replaced at least once a year. There is an enormous variation between species in the number of feathers, for example a whistling swan has over 25,000 contour feathers, whereas a ruby-throated hummingbird has less than 950.

feces
Remains of food and other waste material eliminated from the digestive tract of animals by way of the anus. Feces consist of quantities of fibrous material, bacteria and other microorganisms, rubbed-off lining of the digestive tract, bile fluids, undigested food, minerals, and water.

fecundity
The rate at which an organism reproduces, as distinct from its ability to reproduce (fertility). In vertebrates, it is usually measured as the number of offspring produced by a female each year.

femur
The thigh bone in humans and also the upper bone in the hind limb of a four-limbed vertebrate.

fermentation
The breakdown of sugars by bacteria and yeasts using a method of respiration without oxygen (anaerobic). Fermentation processes have long been used in baking bread, making beer and wine, and producing cheese, yogurt, soy sauce, and many other foodstuffs.

fern
Any of a group of plants of the order Filicales, related to horsetails and club mosses. Ferns are spore-bearing, not flowering, plants and most are perennial, spreading by slow-growing roots. The leaves, known as fronds, vary widely in size and shape. Some taller types, such as tree ferns, grow in the tropics. There are over seven thousand species.

fertility
An organism's ability to reproduce, as distinct from the rate at which it reproduces (fecundity). Individuals become infertile (unable to reproduce) when they cannot generate gametes (eggs or sperm) or when their gametes cannot yield a viable embryo after fertilization.

fertilization
In sexual reproduction, the union of two gametes (sex cells, often called egg and sperm) to produce a zygote, which combines the genetic material contributed

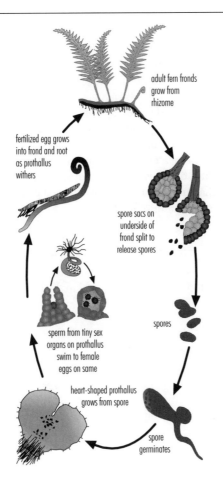

adult fern fronds grow from rhizome

fertilized egg grows into frond and root as prothallus withers

spore sacs on underside of frond split to release spores

sperm from tiny sex organs on prothallus swim to female eggs on same

spores

heart-shaped prothallus grows from spore

spore germinates

fern

by each parent. In self-fertilization the male and female gametes come from the same plant; in cross-fertilization they come from different plants. Self-fertilization rarely occurs in animals; usually even hermaphrodite animals cross-fertilize each other.

fertilizer

Substance containing some or all of a range of about twenty chemical elements necessary for healthy plant growth, used to compensate for the deficiencies of poor or depleted soil. Fertilizers may be organic, for example farmyard manure, composts, bonemeal, blood, and fishmeal; or inorganic, in the form of compounds, mainly of nitrogen, phosphate, and potash, which have been used on a very much increased scale since 1945. Compounds of nitrogen and phosphorus are of particular importance.

fetus

Stage in mammalian embryo development. The human embryo is usually termed a *fetus* after the eighth week of development, when the limbs and external features of the head are recognizable.

fiber, dietary; or roughage

Plant material that cannot be digested by human digestive enzymes; it consists largely of cellulose, a carbohydrate found in plant cell walls. Fiber adds bulk to the gut contents, assisting the muscular contractions that force food along the intestine. A diet low in fiber causes constipation and is believed to increase the risk of developing diverticulitis, diabetes, gall-bladder disease, and cancer of the large bowel—conditions that are rare in nonindustrialized countries, where the diet contains a high proportion of unrefined cereals.

fibrin

Insoluble protein involved in blood clotting. When an injury occurs, fibrin is deposited around the wound in the form of a mesh, which dries and hardens, so that bleeding stops. Fibrin is developed in the blood from a soluble protein, fibrinogen.

fibula

The rear lower bone in the hind leg of a vertebrate. It is paired and often fused with a smaller front bone, the tibia.

fin

In aquatic animals, flattened extension from the body that aids balance and propulsion through the water.

fish

Aquatic vertebrate that uses gills to obtain oxygen from fresh or sea water. There are three main groups: the bony fishes or Osteichthyes (goldfish, cod, tuna); the cartilaginous fishes or Chondrichthyes (sharks, rays); and the jawless fishes or Agnatha (hagfishes, lampreys). Fishes of some form are found in virtually every body of water in the world except for the very salty water of the Dead Sea and some of the hot larval springs. Of the thirty thousand fish species, approximately twenty-five hundred are freshwater.

fitness

In genetic theory, a measure of the success with which a genetically determined character can spread in future generations. By convention, the normal character is assigned a fitness of one, and variants (determined by other alleles) are then assigned fitness values relative to this. Those with fitness greater than one will spread more rapidly and will ultimately replace the normal allele; those with fitness less than one will gradually die out.

flaccidity

In botany, the loss of rigidity (turgor) in plant cells, caused by loss of water from the central vacuole so that the cytoplasm no longer pushes against the cellulose cell wall. If this condition occurs throughout the plant then wilting is seen.

flagellum

Small hairlike organ on the surface of certain cells. Flagella are the motile organs of certain protozoa and single-celled algae, and of the sperm cells of higher animals. Unlike cilia, flagella usually occur singly or in pairs; they are also longer and have a more complex whiplike action.

flatworm

Invertebrate of the phylum Platyhelminthes. Some are free-living, but many are parasitic (for example, tapeworms and flukes). The body is simple and bilaterally symmetrical, with one opening to the intestine. Many are hermaphroditic (with both male and female sex organs) and practice self-fertilization.

flexor

Any muscle that bends a limb. Flexors usually work in opposition to other muscles, the extensors, an arrangement known as antagonistic.

flocculation

In soils, the artificially induced coupling together of particles to improve aeration and drainage. Clay soils, which have very tiny particles and are difficult to work, are often treated in this way. The method involves adding more lime to the soil.

floral diagram

Diagram showing the arrangement and number of parts in a flower, drawn in cross section. An ovary is drawn in the center, surrounded by representations of the other floral parts, indicating the position of each at its base. If any parts such as the petals or sepals are fused, this is also indicated. Floral diagrams allow the structure of different flowers to be compared and are usually shown with the floral formula.

floral formula

Symbolic representation of the structure of a flower. Each kind of floral part is represented by a letter (K for calyx, C for corolla, P for perianth, A for androecium, G for gynoecium) and a number to indicate the quantity of the part present, for example, C5 for a flower with five petals. The number is in brackets if the parts are fused. If the parts are arranged in distinct whorls within the flower, this is shown by two separate figures, such as A5 + 5, indicating two whorls of five stamens each.

floret

Small flower, usually making up part of a larger, composite flower head. There are often two different types present on one flower head: disc florets in the central area, and ray florets around the edge, which usually have a single petal known as the ligule. In the common daisy, for example, the disc florets are yellow, while the ligules are white.

flower

The reproductive unit of an angiosperm or flowering plant, typically consisting of four whorls of modified leaves: sepals, petals, stamens, and carpels. These are borne on a central axis or receptacle. The many variations in size, color, number, and arrangement of parts are closely related to the method of pollination. Flowers adapted for wind pollination typically have reduced or absent petals and sepals and long, feathery stigmas that hang outside the flower to trap airborne pollen. In contrast, the petals of insect-pollinated flowers are usually conspicuous and brightly colored.

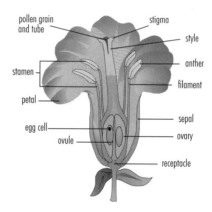

flowering plant

Term generally used for angiosperms, which bear flowers with various parts, including sepals, petals, stamens, and carpels. Sometimes the term is used more broadly, to include both angiosperms and gymnosperms, in which case the cones of conifers and cycads are referred to as "flowers." Usually, however, the angiosperms and gymnosperms are referred to collectively as seed plants, or spermatophytes.

fluorescence microscopy

Technique for examining samples under a microscope without slicing them into thin sections. Instead, fluorescent dyes are introduced into the tissue and used as a light source for imaging purposes. Fluorescent dyes can also be bonded to monoclonal antibodies and used to highlight areas where particular cell proteins occur.

folic acid

A vitamin of the B complex. It is found in liver and green leafy vegetables, and is also synthesized by the intestinal bacteria. It is essential for growth and plays many other roles in the body. Lack of folic acid causes anemia because it is necessary for the synthesis of nucleic acids and the formation of red blood cells.

follicle

In botany, a dry, usually many-seeded fruit that splits along one side only to release the seeds within. It is derived from a single carpel. Examples include the fruits of the larkspurs *Delphinium* and columbine *Aquilegia*. It differs from a pod, which always splits open (dehisces) along both sides.

follicle

A small group of cells that surround and nourish a structure such as a hair (hair follicle) or a cell such as an egg (Graafian follicle).

follicle-stimulating hormone

FSH, a hormone produced by the pituitary gland. It affects the ovaries in women, stimulating the production of an egg cell. Luteinizing hormone is needed to complete the process. In men, FSH stimulates the testes to produce sperm. It is used to treat some forms of infertility.

food

Anything eaten by human beings and other animals and plants to sustain life and health. The building blocks of food are nutrients, and humans can use the following nutrients: (1) carbohydrates, as starches found in bread, potatoes, and pasta; as simple sugars in sucrose and honey; as fibers in cereals, fruit, and vegetables; (2) proteins as from nuts, fish, meat, eggs, milk, and some vegetables; (3) fats as found in most animal products (meat, lard, dairy products, fish), also in margarine, nuts and seeds, olives, and edible oils; (4) vitamins, found in a wide variety of foods, except for vitamin B_{12}, which is found mainly in foods of animal origin; (5) minerals, found in a wide variety of foods (for example, calcium from milk and broccoli, iodine from seafood, and iron from liver and green vegetables); (6) water, ubiquitous in nature; and (7) alcohol, found in fermented distilled beverages, from 40 percent in spirits to 0.01 percent in low-alcohol lagers and beers.

food chain

In ecology, a sequence showing the feeding relationships among organisms in a particular ecosystem. Each organism depends on the next lowest member of the chain for its food. A pyramid of numbers can be used to show the reduction in food energy at each step up the food chain.

food irradiation

The exposure of food to low-level irradiation to kill microorganisms; a technique used in food technology. Irradiation is highly effective and does not make the food any more radioactive than it is naturally. Irradiated food is used for astronauts and immunocompromised patients in hospitals. Some vitamins are partially destroyed, such as vitamin C, and it would be unwise to eat only irradiated fruit and vegetables.

food test

Any of several types of a simple test, easily performed in the laboratory, used to identify the main classes of food: *Starch–iodine test:* food is ground up in distilled water and iodine is added. A dense black color indicates that starch is present. *Sugar–Benedict's test:* food is ground up in distilled water and placed in a test tube with Benedict's reagent. The tube is then heated in a boiling water bath. If glucose is present the color changes from blue to brick-red. *Protein–Biuret test:* food is ground up in distilled water and a mixture of copper (II) sulfate and sodium hydroxide is added. If protein is present a mauve color is seen.

forensic science

The use of scientific techniques to solve criminal cases. A multidisciplinary field embracing chemistry, physics, botany, zoology, and medicine, forensic science includes the identification of human bodies and trace evidence. Ballistics (the study of projectiles, such as bullets), another traditional forensic field, makes use of such tools as the comparison microscope and the electron microscope.

forest

Area where trees have grown naturally for centuries, instead of being logged at maturity (about 150 to 200 years). A natural, or old-growth, forest has a multi-story canopy and includes young and very old trees (this gives the canopy its range of heights). There are also fallen trees contributing to the very complex ecosystem, which may support more than 150 species of mammals and many thousands of species of insects. Globally forest is estimated to have covered around 68 million sq km/26.25 million sq mi during prehistoric times. By the late 1990s, this is believed to have been reduced by half to 34.1 million sq km/ 13.2 million sq mi.

freeze-drying

Method of preserving food. The product to be dried is frozen and then put in a vacuum chamber that forces out the ice as water vapor, a process known as sublimation.

frond

Large leaf or leaflike structure; in ferns it is often pinnately divided. The term is also applied to the leaves of palms and less commonly to the plant bodies of certain seaweeds, liverworts, and lichens.

fruit

From Latin *frui,* meaning "to enjoy," in botany, the ripened ovary in flowering plants that develops from one or more seeds or carpels and encloses one or more seeds. Its function is to protect the seeds during their development and to aid in their dispersal. Fruits are often edible, sweet, juicy, and colorful. When eaten they provide vitamins, minerals, and enzymes, but little protein. Most fruits are borne by perennial plants.

fungus

Plural *fungi,* any of a unique group of organisms of the kingdom Fungi that includes molds, yeasts, rusts, smuts, mildews, mushrooms, and toadstools. About fifty thousand species have been identified. They are not considered to be plants for three main reasons: they have no leaves or roots; they contain no chlorophyll (green coloring) and are therefore unable to make their own food by photosynthesis; and they reproduce by spores. Some fungi are edible but many are highly poisonous; they often cause damage and sometimes disease to the organic matter they live and feed on, but some fungi are exploited in the production of food and drink (for example, yeasts in baking and brewing) and in medicine (for example, penicillin).

fur

The hair of certain animals. Fur is an excellent insulating material and so has been used as clothing. This is, however, vociferously criticized by many groups on humane grounds, as the methods of breeding or trapping animals are often cruel. Mink, chinchilla, and sable are among the most valuable, the wild furs being finer than the farmed. Fur such as mink is made up of a soft, thick, insulating layer called underfur and a top layer of longer, lustrous guard hairs.

gall

Abnormal outgrowth on a plant that develops as a result of attack by insects or, less commonly, by bacteria, fungi, mites, or nematodes. The attack causes an increase in the number of cells or an enlargement of existing cells in the plant. Gall-forming insects generally pass the early stages of their life inside the gall. Gall wasps are responsible for the conspicuous bud galls forming on oak trees, 2.5–4 cm/1–1.5 in across, known as "oak apples." The organisms that cause galls are host-specific. Thus, for example, gall wasps tend to parasitize oaks, and sawflies willows.

gall bladder

Small muscular sac, part of the digestive system of most, but not all, vertebrates. In humans, it is situated on the underside of the liver and connected to the small intestine by the bile duct. It stores bile from the liver.

gamete

Cell that functions in sexual reproduction by merging with another gamete to form a zygote. Examples of gametes include sperm and egg cells. In most organisms, the gametes are haploid (they contain half the number of chromosomes of the parent), owing to reduction division or meiosis.

gametophyte

The haploid generation in the life cycle of a plant that produces gametes.

gamma-interferon

See **interferon.**

ganglion
Plural *ganglia,* solid cluster of nervous tissue containing many cell bodies and synapses, usually enclosed in a tissue sheath; found in invertebrates and vertebrates.

gas exchange
Movement of gases between an organism and the atmosphere, principally oxygen and carbon dioxide. All aerobic organisms (most animals and plants) take in oxygen in order to burn food and manufacture adenosine triphosphate (ATP). The resultant oxidation reactions release carbon dioxide as a waste product to be passed out into the environment. Green plants also absorb carbon dioxide during photosynthesis and release oxygen as a waste product.

gemma
Plural *gemmae,* unit of vegetative reproduction, consisting of a small group of undifferentiated green cells. Gemmae are found in certain mosses and liverworts, forming on the surface of the plant, often in cup-shaped structures, or gemmae cups. Gemmae are dispersed by splashes of rain and can then develop into new plants. In many species, gemmation is more common than reproduction by spores.

gene
Unit of inherited material, encoded by a strand of DNA and transcribed by RNA. In higher organisms, genes are located on the chromosomes. A gene consistently affects a particular character in an individual—for example, the gene for eye color. Also termed a Mendelian gene, after Austrian biologist Gregor Mendel, it occurs at a particular point, or locus, on a particular chromosome and may have several variants, or alleles, each specifying a particular form of that character— for example, the alleles for blue or brown eyes. Some alleles show dominance. These mask the effect of other alleles, known as recessive.

gene amplification
Technique by which selected DNA from a single cell can be duplicated indefinitely until there is a sufficient amount to analyze by conventional genetic techniques.

gene bank
Collection of seeds or other forms of genetic material, such as tubers, spores, bacterial or yeast cultures, live animals and plants, frozen sperm and eggs, or frozen embryos. These are stored for possible future use in agriculture, plant and animal breeding, or in medicine, genetic engineering, or the restocking of wild habitats where species have become extinct. Gene banks will be increasingly used as the rate of extinction increases, depleting the earth's genetic variety (biodiversity).

gene pool
Total sum of alleles (variants of genes) possessed by all the members of a given population or species alive at a particular time.

gene therapy
Medical technique for curing or alleviating inherited diseases or defects; certain infections, and several kinds of cancer in which affected cells from a sufferer would be removed from the body, the DNA repaired in the laboratory (genetic engineering), and the functioning cells reintroduced. A genetically engineered gene was used to treat a patient for the first time in 1990.

genetic code
The way in which instructions for building proteins, the basic structural molecules of living matter, are "written" in the genetic material DNA. This relationship between the sequence of bases (the subunits in a DNA molecule) and the sequence of amino acids (the subunits of a protein molecule) is the basis of heredity. The code employs codons of three bases each; it is the same in almost all organisms, except for a few minor differences recently discovered in some protozoa.

genetic engineering
Deliberate manipulation of genetic material by biochemical techniques. It is often achieved by the introduction of new DNA, usually by means of a virus or plasmid. This can be for pure research, gene therapy, or to breed functionally specific plants, animals, or bacteria. These organisms with a foreign gene added are said to be transgenic. At the beginning of 1995, more than sixty plant species had been genetically engineered, and nearly three thousand transgenic crops had been field-tested.

genetic fingerprinting, or genetic profiling
Technique used for determining the pattern of certain parts of the genetic material DNA that is unique to each individual. Like conventional fingerprinting, it can accurately distinguish humans from one another, with the exception of identical siblings from a multiple birth. It can be applied to as little material as a single cell.

genetics
Branch of biology concerned with the study of heredity and variation; it attempts to explain how characteristics of living organisms are passed on from one generation to the next. The science of genetics is based on the work of Austrian biologist Gregor Mendel, whose experiments with the cross-breeding (hybridization) of peas showed that the inheritance of characteristics and traits takes place by means of discrete "particles" (genes). These are present in the cells of all organisms and are now recognized as being the basic units of heredity. All organisms possess genotypes (sets of variable genes) and phenotypes (characteristics

produced by certain genes). Modern geneticists investigate the structure, function, and transmission of genes.

genitalia
Reproductive organs of sexually reproducing animals, particularly the external/visible organs of mammals: in males, the penis and the scrotum, which contains the testes, and in females, the clitoris and vulva.

genome
The full complement of genes carried by a single (haploid) set of chromosomes. The term may be applied to the genetic information carried by an individual or to the range of genes found in a given species. The human genome is made up of seventy-five thousand genes.

genotype
The particular set of alleles (variants of genes) possessed by a given organism. The term is usually used in conjunction with *phenotype,* which is the product of the genotype and all environmental effects.

genus
Plural *genera,* group of species with many characteristics in common. Thus all doglike species (including dogs, wolves, and jackals) belong to the genus *Canis* (Latin "dog"). Species of the same genus are thought to be descended from a common ancestor species. Related genera are grouped into families.

germ
Colloquial term for a microorganism that causes disease, such as certain bacteria and viruses. Formerly, it was also used to mean something capable of developing into a complete organism (such as a fertilized egg, or the embryo of a seed).

germination
In botany, the initial stages of growth in a seed, spore, or pollen grain. Seeds germinate when they are exposed to favorable external conditions of moisture, light, and temperature, and when any factors causing dormancy have been removed.

germ layer
In embryology, a layer of cells that can be distinguished during the development of a fertilized egg. Most animals have three such layers: the inner, middle, and outer. These differentiate to form the various body tissues.

gestation
In all mammals except the monotremes (platypus and spiny anteaters), the period from the time of implantation of the embryo in the uterus to birth. This period varies among species; in humans it is about 266 days, in elephants 18 to 22

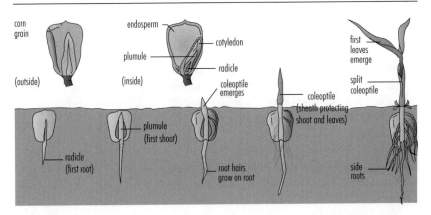

germination

months, in cats about 60 days, and in some species of marsupial (such as opossum) as short as 12 days.

gibberellin
Plant growth substance that promotes stem growth and may also affect the breaking of dormancy in certain buds and seeds and the induction of flowering. Application of gibberellin can stimulate the stems of dwarf plants to additional growth, delay the aging process in leaves, and promote the production of seedless fruit (parthenocarpy).

gill
The main respiratory organ of most fishes and immature amphibians, and of many aquatic invertebrates. In all types, water passes over the gills, and oxygen diffuses across the gill membranes into the circulatory system, while carbon dioxide passes from the system out into the water.

gizzard
Muscular grinding organ of the digestive tract, below the crop of birds, earthworms, and some insects, and forming part of the stomach. The gizzard of birds is lined with a hardened horny layer of the protein keratin, preventing damage to the muscle layer during the grinding process. Most birds swallow sharp grit, which aids maceration of food in the gizzard.

gland
Specialized organ of the body that manufactures and secretes enzymes, hormones, or other chemicals. In animals, glands vary in size from small (for example, tear glands) to large (for example, the pancreas), but in plants they are always small, and may consist of a single cell. Some glands discharge their products internally;

endocrine glands, and others, exocrine glands, externally. Lymph nodes are sometimes wrongly called glands.

glomerulus

In the kidney, the cluster of blood capillaries at the threshold of the renal tubule, or nephron, responsible for filtering out the fluid that passes down the tubules and ultimately becomes urine. The human kidney has approximately one million tubules, each possessing its own glomerulus.

glucagon

A hormone secreted by the alpha cells of the islets of Langerhans in the pancreas, which increases the concentration of glucose in the blood by promoting the breakdown of glycogen in the liver. Secretion occurs in response to a lowering of blood glucose concentrations.

glucose, or **dextrose** or **grape sugar**

Sugar ($C_6H_{12}O_6$) present in the blood and manufactured by green plants during photosynthesis. The respiration reactions inside cells involve the oxidation of glucose to produce adenosine triphosphate (ATP), the "energy molecule" used to drive many of the body's biochemical reactions.

glycogen

Polymer (a polysaccharide) of the sugar glucose made and retained in the liver as a carbohydrate store, for which reason it is sometimes called animal starch. It is a source of energy when needed by muscles, where it is converted back into glucose by the hormone insulin and metabolized.

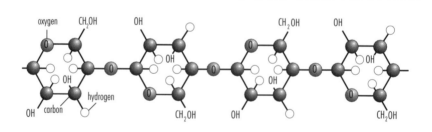

goblet cell

Cup-shaped cell present in the epithelium of the respiratory and gastrointestinal tracts. Goblet cells secrete mucin, the main constituent of mucous, which lubricates the mucous membranes of these tracts.

Golgi apparatus, or **Golgi body**

Stack of flattened membranous sacs found in the cells of eukaryotes. Many molecules travel through the Golgi apparatus on their way to other organelles or to the endoplasmic reticulum. Some are modified or assembled inside the sacs. The Golgi apparatus is named for the Italian physician Camillo Golgi (1843[?]–1926).

gonad
The part of an animal's body that produces the sperm or egg cells (ova) required for sexual reproduction. The sperm-producing gonad is called a testis, and the egg-producing gonad is called an ovary.

Graafian follicle
Fluid-filled capsule that surrounds and protects the developing egg cell inside the ovary during the menstrual cycle. After the egg cell has been released, the follicle remains and is known as a corpus luteum.

grafting
In medicine, the operation by which an organ or other living tissue is removed from one organism and transplanted into the same or a different organism. In horticulture, it is a technique widely used for propagating plants, especially woody species. A bud or shoot on one plant, termed the scion, is inserted into another, the stock, so that they continue growing together, the tissues combining at the point of union. In this way some of the advantages of both plants are obtained.

grape sugar
See glucose.

grass
Any of a very large family of plants of the family Gramineae, many of which are economically important because they provide grazing for animals and food for humans in the form of cereals. There are about nine thousand grass species distributed worldwide except in the Arctic regions. Most are perennial, with long, narrow leaves and jointed, hollow stems; flowers with both male and female reproductive organs are borne on spikelets; the fruits are grainlike. Included in the family are bluegrass, wheat, rye, corn, sugarcane, and bamboo.

gray matter
Those parts of the brain and spinal cord that are made up of interconnected and tightly packed nerve cell nucleuses. The outer layers of the cerebellum contain most of the gray matter in the brain. It is the region of the brain that is responsible for advanced mental functions. Grey matter also constitutes the inner core of the spinal cord. This is in contrast to white matter, which is made of the axons of nerve cells.

grooming
The use by an animal of teeth, tongue, feet, or beak to clean fur or feathers. Grooming also helps to spread essential oils for waterproofing. In many social species, notably monkeys and apes, grooming of other individuals is used to reinforce social relationships.

growth rings
See **annual rings.**

grub
Legless larval stages of Coleoptera (beetles) and Hymenoptera (bees, ants, and wasps).

guard cell
In plants, a specialized cell on the undersurface of leaves for controlling gas exchange and water loss. Guard cells occur in pairs and are shaped so that a pore, or stomata, exists between them. They can change shape with the result that the pore disappears. During warm weather, when a plant is in danger of losing excessive water, the guard cells close, cutting down evaporation from the interior of the leaf.

gum, mammal
In mammals, the soft tissues surrounding the base of the teeth. Gums are liable to inflammation (gingivitis) or to infection by microbes from food deposits (periodontal disease).

gum, plant
In botany, complex polysaccharides (carbohydrates) formed by many plants and trees, particularly by those from dry regions. They form four main groups: plant exudates (gum arabic); marine plant extracts (agar); seed extracts; and fruit and vegetable extracts. Some are made synthetically.

gut
See **alimentary canal.**

guttation
Secretion of water onto the surface of leaves through specialized pores, or hydathodes. The process occurs most frequently during conditions of high humidity when the rate of transpiration is low. Drops of water found on grass in early morning are often the result of guttation, rather than dew. Sometimes the water contains minerals in solution, such as calcium, which leaves a white crust on the leaf surface as it dries.

gymnosperm
From Greek, meaning "naked seed," in botany, any plant whose seeds are exposed, as opposed to the structurally more advanced angiosperms, the seeds of which are inside an ovary. The group includes conifers and related plants such as cycads and ginkgos, whose seeds develop in cones. Fossil gymnosperms have been found in rocks about 350 million years old.

gynoecium, or gynecium
Collective term for the female reproductive organs of a flower, consisting of one or more carpels, either free or fused together.

habitat

Localized environment in which an organism lives and which provides for all (or almost all) of its needs. The diversity of habitats found within the earth's ecosystem is enormous, and they are changing all the time. Many can be considered inorganic or physical; for example, the Arctic ice cap, a cave, or a cliff face. Others are more complex; for instance, a woodland or a forest floor. Some habitats are so precise that they are called *microhabitats,* such as the area under a stone where a particular type of insect lives. Most habitats provide a home for many species.

hair

Fine filament growing from mammalian skin. Each hair grows from a pit-shaped follicle embedded in the second layer of the skin, the dermis. It consists of dead cells impregnated with the protein keratin.

halophyte

Plant adapted to live where there is a high concentration of salt in the soil, for example, in salt marshes and mud flats.

hapaxanthic

See **monocarpic.**

haploid

Having a single set of chromosomes in each cell. Most higher organisms are diploid—that is, they have two sets—but their gametes (sex cells) are haploid. Some plants, such as mosses, liverworts, and many seaweeds, are haploid, and male honey bees are haploid because they develop from eggs that have not been fertilized.

haustorium

Plural *haustoria,* specialized organ produced by a parasitic plant or fungus that penetrates the cells of its host to absorb nutrients. It may be either an outgrowth of hyphae, as in the case of parasitic fungi, or of the stems of flowering parasitic plants, as in dodders *(Cuscuta).* The suckerlike haustoria of a dodder penetrate the vascular tissue of the host plant without killing the cells.

heart

Muscular organ that rhythmically contracts to force blood around the body of an animal with a circulatory system. Annelid worms and some other invertebrates have simple hearts consisting of thickened sections of main blood vessels that pulse regularly. An earthworm has ten such hearts. Vertebrates have one heart. A fish heart has two chambers—the thin-walled atrium (once called the auricle) that expands to receive blood, and the thick-walled ventricle that pumps it out. Amphibians and most reptiles have two atria and one ventricle; birds and mammals have two atria and two ventricles. The beating of the heart is controlled by the autonomic nervous system and an internal control center or pacemaker, the sinoatrial node.

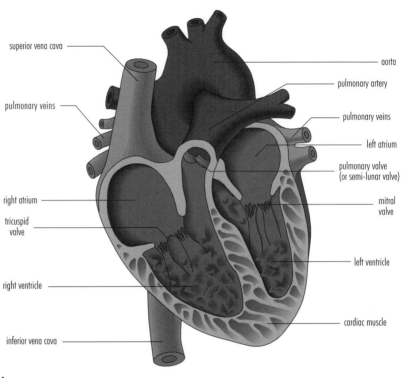

superior vena cava

aorta

pulmonary artery

pulmonary veins

pulmonary veins

left atrium

pulmonary valve
(or semi-lunar valve)

right atrium

mitral
valve

tricuspid
valve

left ventricle

right ventricle

cardiac muscle

inferior vena cava

heart

heartbeat

The regular contraction and relaxation of the heart, and the accompanying sounds. As blood passes through the heart a double beat is heard. The first is produced by the sudden closure of the valves between the atria and the ventricles. The second, slightly delayed sound, is caused by the closure of the valves found at the entrance to the major arteries leaving the heart. Diseased valves may make unusual sounds, known as heart murmurs.

hematocrit

The percentage of whole blood volume represented by the red blood cells.

hemoglobin

Protein used by all vertebrates and some invertebrates for oxygen transportation because the two substances combine reversibly. In vertebrates it occurs in red blood cells (erythrocytes), giving them their color.

hemolymph

Circulatory fluid of those mollusks and insects that have an "open" circulatory system. Hemolymph contains water, amino acids, sugars, salts, and white cells like those of blood. Circulated by a pulsating heart, its main function is to transport digestive and excretory products around the body. In mollusks, it also transports oxygen and carbon dioxide.

hemolysis
Destruction of red blood cells. Aged cells are constantly being lysed (broken down), but increased wastage of red cells is seen in some infections and blood disorders. It may result in jaundice (through the release of too much hemoglobin) and in anemia.

herb
Any plant (usually a flowering plant) tasting sweet, bitter, aromatic, or pungent, used in cooking, medicine, or perfumery; technically, a herb is any plant in which the aerial parts do not remain above ground at the end of the growing season.

herbaceous plant
Plant with very little or no wood, dying back at the end of every summer. The herbaceous perennials survive winters as underground storage organs such as bulbs and tubers.

herbarium
Collection of dried, pressed plants used as an aid to identification of unknown plants and by taxonomists in the classification of plants. The plant specimens are accompanied by information, such as the date and place of collection, by whom collected, details of habitat, flower color, and local names.

herbicide
Any chemical used to destroy plants or check their growth.

herbivore
Animal that feeds on green plants (or photosynthetic single-celled organisms) or their products, including seeds, fruit, and nectar. The most numerous type of herbivore is thought to be the zooplankton, tiny invertebrates in the surface waters of the oceans that feed on small photosynthetic algae. Herbivores are more numerous than other animals because their food is the most abundant. They form a vital link in the food chain between plants and carnivores.

heredity
The transmission of traits from parent to offspring.

hermaphrodite
Organism that has both male and female sex organs. Hermaphroditism is the norm in such species as earthworms and snails and is common in flowering plants. Cross-fertilization is the rule among hermaphrodites, with the parents function-

ing as male and female simultaneously, or as one or the other sex at different stages in their development. Human hermaphrodites are extremely rare.

heterostyly

In botany, having styles of different lengths. Certain flowers, such as primroses *(Primula vulgaris),* have different-sized anthers and styles to ensure cross-fertilization (through pollination) by visiting insects.

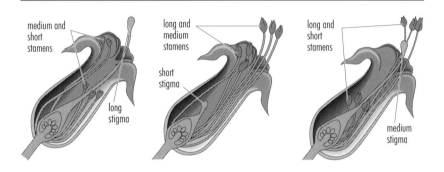

heterotroph

Any living organism that obtains its energy from organic substances produced by other organisms. All animals and fungi are heterotrophs, and they include herbivores, carnivores, and saprotrophs (those that feed on dead animal and plant material).

heterozygous

In a living organism, having two different alleles for a given trait. In homozygous organisms, by contrast, both chromosomes carry the same allele. In an outbreeding population an individual organism will generally be heterozygous for some genes but homozygous for others.

hibernation

State of dormancy in which certain animals spend the winter. It is associated with a dramatic reduction in all metabolic processes, including body temperature, breathing, and heart rate. It is a fallacy that such animals sleep nonstop throughout the winter.

high-yield variety

Crop that has been specially bred or selected to produce more than the natural varieties of the same species. During the 1950s and 1960s, new strains of wheat and corn were developed to reduce the food shortages in poor countries (the Green Revolution). Later, IR8, a new variety of rice that increased yields by up to six times, was developed in the Philippines. Strains of crops resistant to drought and disease were also developed. High-yield varieties require large amounts of expensive artificial fertilizers and sometimes pesticides for best results.

hinge joint
In vertebrates, a joint where movement occurs in one plane only. Examples are the elbow and knee, which are controlled by pairs of muscles, the flexors and extensors.

histology
Study of plant and animal tissue by visual examination, usually with a microscope.

HIV
Abbreviation for human immunodeficiency virus, the infectious agent that is believed to cause AIDS (Acquired Immune Deficiency Syndrome). It was first discovered in 1983 by Luc Montagnier of the Pasteur Institute in Paris, who called it lymphocyte-associated virus (LAV). Independently, U.S. scientist Robert Gallo of the National Cancer Institute in Bethesda, Maryland, claimed its discovery in 1984 and named it human T-lymphocytotrophic virus 3 (HTLV-III).

homeostasis
Maintenance of a constant environment around living cells, particularly with regard to pH, salt concentration, temperature, and blood sugar levels. Stable conditions are important for the efficient functioning of the enzyme reactions within the cells. In humans, homeostasis in the blood (which provides fluid for all tissues) is ensured by several organs. The kidneys regulate pH, urea, and water concentration. The lungs regulate oxygen and carbon dioxide. Temperature is regulated by the liver and the skin. Glucose levels in the blood are regulated by the liver and the pancreas.

homeothermy
Maintenance of a constant body temperature in endothermic (warm-blooded) animals by the use of chemical processes to compensate for heat loss or gain when external temperatures change. Such processes include generation of heat by the breakdown of food and the contraction of muscles, and loss of heat by sweating, panting, and other means.

homologous
A term describing an organ or structure possessed by members of different taxonomic groups (for example, species, genera, families, orders) that originally derived from the same structure in a common ancestor. The wing of a bat, the arm of a monkey, and the flipper of a seal are homologous because they all derive from the forelimb of an ancestral mammal.

homozygous

In a living organism, having two identical alleles for a given trait. Individuals homozygous for a trait always breed true; that is, they produce offspring that resemble them in appearance when bred with a genetically similar individual; inbred varieties or species are homozygous for almost all traits. Recessive alleles are only expressed in the homozygous condition.

honey guide

In botany, line or spot on the petals of a flower that indicates to pollinating insects the position of the nectaries within the flower. The orange dot on the lower lip of the toadflax flower *(Linaria vulgaris)* is an example. Sometimes the markings reflect only ultraviolet light, which can be seen by many insects although it is not visible to the human eye.

hormone

Chemical secretion of the ductless endocrine glands and specialized nerve cells concerned with control of body functions. The major glands are the thyroid, parathyroid, pituitary, adrenal, pancreas, ovary, and testis. There are also hormone-secreting cells in the kidney, liver, gastrointestinal tract, thymus (in the neck), pineal (in the brain), and placenta. Hormones bring about changes in the functions of various organs according to the body's requirements. The hypothalamus, which adjoins the pituitary gland at the base of the brain, is a control center for overall coordination of hormone secretion; the thyroid hormones determine the rate of general body chemistry; the adrenal hormones prepare the organism during stress for "fight or flight"; and the sexual hormones such as estrogen and testosterone govern reproductive functions.

hornwort

Nonvascular plant (with no "veins" to carry water and food) related to the liverworts and mosses. Hornworts are found in warm climates, growing on moist, shaded soil (class Anthocerotae, order Bryophyta). The name is also given to a group of aquatic flowering plants that are found in slow-moving water. They have whorls of finely divided leaves and may grow up to 2 m/7 ft long (genus *Ceratophyllum,* family Ceratophyllaceae).

horticulture

Art and science of growing flowers, fruit, and vegetables. Horticulture is practiced in gardens and orchards, along with millions of acres of land devoted to vegetable farming. Some areas, like California, have specialized in horticulture because they have the mild climate and light fertile soil most suited to these crops.

host

An organism that is parasitized by another. In commensalism, the partner that does not benefit may also be called the host.

human body

The physical structure of the human being. It develops from the single cell of the fertilized ovum, is born at forty weeks, and usually reaches sexual maturity between eleven and eighteen years of age. The bony framework (skeleton) consists of more than two hundred bones, more than half of which are in the hands and feet. Bones are held together by joints, some of which allow movement. The circulatory system supplies muscles and organs with blood, which provides oxygen and food and removes carbon dioxide and other waste products. Body functions are controlled by the nervous system and hormones. In the upper part of the trunk is the thorax, which contains the lungs and heart. Below this is the abdomen, containing the digestive system (stomach and intestines); the liver, spleen, and pancreas; the urinary system (kidneys, ureters, and bladder); and, in women, the reproductive organs (ovaries, uterus, and vagina). In men, the prostate gland and seminal vesicles of the reproductive system are situated in the abdomen, the testes being in the scrotum, which, with the penis, is suspended in front of and below the abdomen. The bladder empties through a small channel (urethra); in the female this opens in the upper end of the vulval cleft, which also contains the opening of the vagina, or birth canal; in the male, the urethra is continued into the penis. In both sexes, the lower bowel terminates in the anus, a ring of strong muscle situated between the buttocks.

Human Genome Project

Research project, begun in 1988, to map the complete nucleotide sequence of human DNA. There are approximately eighty thousand different genes in the human genome, and one gene may contain more than two million nucleotides. The program aims to collect ten to fifteen thousand genetic specimens from 722 ethnic groups whose genetic makeup is to be preserved for future use and study. The knowledge gained is expected to help prevent or treat many crippling and lethal diseases, but there are potential ethical problems associated with knowledge of an individual's genetic makeup, and fears that it will lead to genetic discrimination. Only 3 percent of the genome had been sequenced by mid-1998, but plans were announced to complete a "rough draft" of 95 percent of the genome over the next three years. The target date for sequencing the whole genome is 2005, though a private U.S. company announced plans in 1998 to sequence the entire genome by 2001.

human immunodeficiency virus

See HIV.

human reproduction

An example of sexual reproduction, where the male produces sperm and the female produces eggs. These gametes contain only half the normal number of chromosomes, twenty-three instead of forty-six, so that on fertilization the resulting cell has the correct genetic complement. Fertilization is internal, which increases the chances of conception; unusually for mammals, copulation and pregnancy can occur at any time of the year. Human beings are also remarkable for the length of childhood and for the highly complex systems of parental care found in society. The use of contraception and the development of laboratory methods of insemination and fertilization are issues that make human reproduction more than a merely biological phenomenon.

humerus
The upper bone of the forelimb of tetrapods. In humans, the humerus is the bone above the elbow.

hybrid
Offspring from a cross between individuals of two different species, or two inbred lines within a species. In most cases, hybrids between species are infertile and unable to reproduce sexually. In plants, however, doubling of the chromosomes can restore the fertility of such hybrids.

hydathode
Specialized pore, or less commonly, a hair, through which water is secreted by hydrostatic pressure from the interior of a plant leaf onto the surface. Hydathodes are found on many different plants and are usually situated around the leaf margin at vein endings. Each pore is surrounded by two crescent-shaped cells and resembles an open stoma, but the size of the opening cannot be varied as in a stoma. The process of water secretion through hydathodes is known as guttation.

hydrophily
Type of pollination where the pollen is transported by water. Water-pollinated plants occur in thirty-one genera in eleven different families. They are found in habitats as diverse as rain forests and seasonal desert pools. Pollen is either dispersed underwater or on the water's surface.

hydrophyte
Plant adapted to live in water, or in waterlogged soil.

hydroponics
Cultivation of plants without soil, using specially prepared solutions of mineral salts. Beginning in the 1930s, large crops were grown by hydroponic methods, at first in California but since then in many other parts of the world.

hypha
Plural *hyphae,* delicate, usually branching filament, many of which collectively form the mycelium and fruiting bodies of a fungus. Food molecules and other substances are transported along hyphae by the movement of the cytoplasm, known as "cytoplasmic streaming."

hypogeal
Term used to describe seed germination in which the cotyledons remain below ground. It can refer to fruits that develop underground, such as peanuts *Arachis hypogea*.

hypothalamus
Region of the brain below the cerebrum which regulates rhythmic activity and physiological stability within the body, including water balance and temperature. It regulates the production of the pituitary gland's hormones and controls that part of the nervous system governing the involuntary muscles.

hypoxia
See anoxia.

ileum
Part of the small intestine of the digestive system between the duodenum and the colon that absorbs digested food.

immunity
The protection that organisms have against foreign microorganisms, such as bacteria and viruses, and against cancerous cells. The cells that provide this protection are called white blood cells, or leukocytes, and make up the immune system. They include neutrophils and macrophages, which can engulf invading organisms and other unwanted material, and natural killer cells that destroy cells infected by viruses and cancerous cells. Some of the most important immune cells are the B cells and T cells. Immune cells coordinate their activities by means of chemical messengers or lymphokines, including the antiviral messenger interferon. The lymph nodes play a major role in organizing the immune response.

immunocompromised
Lacking a fully effective immune system. The term is most often used in connection with infections such as AIDS where the virus interferes with the immune response.

immunodeficient
Lacking one or more elements of a working immune system. Immune deficiency is the term generally used for patients who are born with such a defect, while those who acquire such a deficiency later in life are referred to as immunocompromised or immunosuppressed.

immunoglobulin
Human globulin protein that can be separated from blood and administered to confer immediate immunity on the recipient. It participates in the immune reaction as the antibody for a specific antigen (disease-causing agent).

implantation

In mammals, the process by which the developing embryo attaches itself to the wall of the mother's uterus and stimulates the development of the placenta. In humans it occurs six to eight days after ovulation.

imprinting

In ethology, the process whereby a young animal learns to recognize both specific individuals (for example, its mother) and its own species.

inbreeding

In genetics, the mating of closely related individuals. It is considered undesirable because it increases the risk that offspring will inherit copies of rare deleterious recessive alleles (genes) from both parents and so suffer from disabilities.

incisor

Sharp tooth at the front of the mammalian mouth. Incisors are used for biting or nibbling, as when a rabbit or a sheep eats grass. Rodents, such as rats and squirrels, have large continually growing incisors, adapted for gnawing. The elephant tusk is a greatly enlarged incisor. In humans, the incisors are the four teeth at the front center of each jaw.

inclusive fitness

In genetics, the success with which a given variant (or allele) of a gene is passed on to future generations by a particular individual, after additional copies of the allele in the individual's relatives and their offspring have been taken into account.

indicator species

Plant or animal whose presence or absence in an area indicates certain environmental conditions, such as soil type, high levels of pollution, or, in rivers, low levels of dissolved oxygen. Many plants show a preference for either alkaline or acid soil conditions, while certain trees require aluminum, and are found only in soils where it is present. Some lichens are sensitive to sulfur dioxide in the air, and absence of these species indicates atmospheric pollution.

inflorescence

In plants, a branch, or system of branches, bearing two or more individual flowers. Inflorescences can be divided into two main types: cymose (or definite) and racemose (or indefinite). In a cymose inflorescence, the tip of the main axis produces a single flower and subsequent flowers arise on lower side branches, as in forget-me-not *Myosotis* and chickweed *Stellaria;* the oldest flowers are, therefore, found at the tip. A racemose inflorescence has an active growing region at the tip of its main axis and bears flowers along its length, as in hyacinth *Hyacinthus;* the oldest flowers are found near the base or, in cases where the inflorescence is flattened, toward the outside.

ingestion

Process of taking food into the mouth. The method of food capture varies but may involve biting, sucking, or filtering. Many single-celled organisms have a region of their cell wall that acts as a mouth. In these cases surrounding tiny hairs (cilia) sweep food particles together, ready for ingestion.

inhibition, neural

The process in which activity in one nerve cell suppresses activity in another. Neural inhibition in networks of nerve cells leading from sensory organs, or to muscles, plays an important role in allowing an animal to make fine sensory discriminations and to exercise fine control over movements.

inorganic compound

Compound found in organisms that are not typically biological. Water, sodium chloride, and potassium are inorganic compounds because they are widely found outside living cells. The term is also applied to those compounds that do not contain carbon and that are not manufactured by organisms. However, carbon dioxide is considered inorganic, contains carbon, and is manufactured by organisms during respiration.

insect

Any of a vast group of small invertebrate animals of the class Insecta with hard, segmented bodies, three pairs of jointed legs, and, usually, two pairs of wings; they belong among the arthropods and are distributed throughout the world. An insect's body is divided into three segments: head, thorax, and abdomen. On the head is a pair of feelers, or antennae. The legs and wings are attached to the thorax, or middle segment of the body. The abdomen, or end segment of the body, is where food is digested and excreted and where the reproductive organs are located. Insects vary in size from 0.02 cm/0.007 in to 35 cm/13.5 in in length. The world's smallest insect is believed to be a "fairy fly" wasp in the family Mymaridae, with a wingspan of 0.2 mm/0.008 in.

insecticide

Any chemical pesticide used to kill insects. Among the most effective insecticides are synthetic organic chemicals such as DDT and dieldrin, which are chlorinated hydrocarbons. These chemicals, however, have proved persistent in the environment and are also poisonous to all animal life, including humans, and are consequently banned in many countries. Other synthetic insecticides include organic phosphorus compounds such as malathion. Insecticides prepared from plants, such as derris and pyrethrum, are safer to use but need to be applied frequently and carefully.

insectivore

Any animal whose diet is made up largely or exclusively of insects. In particular, the name is applied to mammals of the order Insectivora, which includes the shrews, hedgehogs, moles, and tenrecs.

insectivorous plant

Plant that can capture and digest live prey (normally insects) to obtain nitrogen compounds that are lacking in its usual marshy habitat. Some are passive traps, for example, the pitcher plants *Nepenthes* and *Sarracenia*. One pitcher-plant species has container-traps holding 1.6 l/3.5 pt. of the liquid that "digests" its food, mostly insects but occasionally even rodents. Others, for example, sundews *Drosera*, butterworts *Pinguicula*, and Venus flytraps *Dionaea muscipula*, have an active trapping mechanism. Insectivorous plants have adapted to grow in poor soil conditions where the number of microorganisms recycling nitrogen compounds is very much reduced. In these circumstances other plants cannot gain enough nitrates to grow.

instinct

In ethology, behavior found in all equivalent members of a given species (for example, all the males, or all the females with young) that is presumed to be genetically determined.

insulin

Protein hormone, produced by specialized cells in the islets of Langerhans in the pancreas, that regulates the metabolism (rate of activity) of glucose, fats, and proteins. Insulin was discovered by Canadian physician Frederick Banting and Canadian physiologist Charles Best, who pioneered its use in treating diabetes.

integument

In seed-producing plants, the protective coat surrounding the ovule. In flowering plants there are two, in gymnosperms only one. A small hole at one end, the micropyle, allows a pollen tube to penetrate through to the egg during fertilization.

intercostal

The nerves, blood vessels, and muscles that lie between the ribs.

interferon

Naturally occurring cellular protein that makes up part of the body's defenses against viral disease. Three types (alpha-, beta-, and gamma-) are produced by infected cells and enter the bloodstream and uninfected cells, making them immune to virus attack.

International Union for the Conservation of Nature (IUCN)

Organization established by the United Nations to promote the conservation of wildlife and habitats as part of the national policies of member states.

intersex

Individual that is intermediate between a normal male and a normal female in its appearance (for example, a genetic male that lacks external genitalia and so resembles a female).

interstitial

Undifferentiated tissue that is interspersed with the characteristic tissue of an organ. It is often formed of fibrous tissue and supports the organ. Interstitial fluid refers to the fluid present in small amounts in the tissues of an organ.

intestine

In vertebrates, the digestive tract from the stomach outlet to the anus. The human small intestine is 6 m/20 ft long, 4 cm/1.5 in in diameter, and consists of the duodenum, jejunum, and ileum; the large intestine is 1.5 m/5 ft long, 6 cm/2.5 in in diameter, and includes the cecum, colon, and rectum. Both are muscular tubes comprising an inner lining that secretes alkaline digestive juice, a submucous coat containing fine blood vessels and nerves, a muscular coat, and a serous coat covering all, supported by a strong peritoneum, which carries the blood and lymph vessels, and the nerves. The contents are passed along slowly by peristalsis (waves of involuntary muscular action). The term *intestine* is also applied to the lower digestive tract of invertebrates.

invertebrate

Animal without a backbone. The invertebrates comprise over 95 percent of the million or so existing animal species and include sponges, coelenterates, flatworms, nematodes, annelid worms, arthropods, mollusks, echinoderms, and primitive aquatic chordates, such as sea squirts and lancelets.

in vitro process

Biological experiment or technique carried out in a laboratory, outside the body of a living organism (literally "in glass," for example in a test tube). By contrast, an *in vivo* process takes place within the body of an organism.

in vivo process

Biological experiment or technique carried out within a living organism; by contrast, an *in vitro* process takes place outside the organism, in an artificial environment such as a laboratory.

iris

In anatomy, the colored muscular diaphragm that controls the size of the pupil in the vertebrate eye. It contains radial muscle that increases the pupil diameter and circular muscle that constricts the pupil diameter. Both types of muscle respond involuntarily to light intensity.

islets of Langerhans

Groups of cells within the pancreas responsible for the secretion of the hormone insulin. They are sensitive to the blood sugar, producing more hormone when glucose levels rise.

iteroparity

The repeated production of offspring at intervals throughout the life cycle. It is usually contrasted with semelparity, where each individual reproduces only once during its life.

jaw

One of two bony structures that form the framework of the mouth in all vertebrates except lampreys and hagfishes (the agnathous or jawless vertebrates). They consist of the upper jawbone (maxilla), which is fused to the skull, and the lower jawbone (mandible), which is hinged at each side to the bones of the temple by ligaments.

joint

In any animal with a skeleton, a point of movement or articulation. In vertebrates, it is the point where two bones meet. Some joints allow no motion (the sutures of the skull), others allow a very small motion (the sacroiliac joints in the lower back), but most allow a relatively free motion. Of these, some allow a gliding motion (one vertebra of the spine on another), some have a hinge action (elbow and knee), and others allow motion in all directions (hip and shoulder joints) by means of a ball-and-socket arrangement. The ends of the bones at a moving joint are covered with cartilage for greater elasticity and smoothness, and enclosed in an envelope (capsule) of tough white fibrous tissue lined with a membrane that secretes a lubricating and cushioning synovial fluid. The joint is further strengthened by ligaments. In invertebrates with an exoskeleton, the joints are places where the exoskeleton is replaced by a more flexible outer covering, the arthrodial membrane, which allows the limb (or other body part) to bend at that point.

jugular vein

One of two veins in the necks of vertebrates; they return blood from the head to the superior (or anterior) vena cava and thence to the heart.

karyotype

The set of chromosomes characteristic of a given species. It is described as the number, shape, and size of the chromosomes in a single cell of an organism. In humans, for example, the karyotype consists of forty-six chromosomes, in mice forty, crayfish two hundred, and in fruit flies eight.

keratin

Fibrous protein found in the skin of vertebrates and also in hair, nails, claws, hooves, feathers, and the outer coating of horns.

kernel

The inner, softer part of a nut, or of a seed within a hard shell.

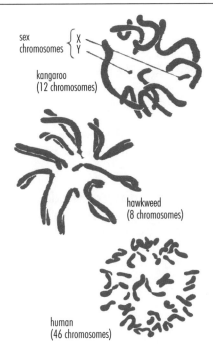

sex
chromosomes { X
            { Y

kangaroo
(12 chromosomes)

hawkweed
(8 chromosomes)

human
(46 chromosomes)

karyotype

kidney

In vertebrates, one of a pair of organs responsible for fluid regulation, excretion of waste products, and maintaining the ionic composition of the blood. The kidneys are situated on the rear wall of the abdomen. Each one consists of a number of long tubules; the outer parts filter the aqueous components of blood, and the inner parts selectively reabsorb vital salts, leaving waste products in the remaining fluid (urine), which is passed through the ureter to the bladder.

kinesis

Plural *kineses*, a nondirectional movement in response to a stimulus; for example, wood lice move faster in drier surroundings. Taxis is a similar pattern of behavior, but there the response is directional.

kingdom

The primary division in biological classification. At one time, only two kingdoms were recognized: animals and plants. Today most biologists prefer a five-kingdom system, even though it still involves grouping together organisms that are probably unrelated. One widely accepted scheme is as follows: Kingdom Animalia (all multicellular animals); Kingdom Plantae (all plants, including seaweeds and other algae, except blue-green); Kingdom Fungi (all fungi, including the unicellular yeasts, but not slime molds); Kingdom Protista or Protoctista (protozoa, diatoms, dinoflagellates, slime molds, and various other lower organisms with eukaryotic cells); and Kingdom Monera (all prokaryotes—the bacteria and cyanobacteria, or blue-green algae). The first four of these kingdoms make up the eukaryotes.

kin selection
The idea that altruism shown to genetic relatives can be worthwhile, because those relatives share some genes with the individual that is behaving altruistically, and may continue to reproduce.

Krebs cycle, or citric acid cycle or tricarboxylic acid cycle
Final part of the chain of biochemical reactions by which organisms break down food using oxygen to release energy (respiration). It takes place within structures called mitochondria in the body's cells, breaking down food molecules in a series of small steps to produce energy-rich molecules of ATP.

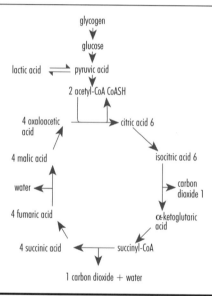

labellum
Lower petal of an orchid flower; it is a different shape from the two lateral petals and gives the orchid its characteristic appearance. The labellum is more elaborate and usually larger than the other petals. It often has distinctive patterning to encourage pollination by insects; sometimes it is extended backward to form a hollow spur containing nectar.

lactation
Secretion of milk in mammals, from the mammary glands. In late pregnancy, the cells lining the lobules inside the mammary glands begin extracting substances from the blood to produce milk. The supply of milk starts shortly after birth with the production of colostrum, a clear fluid consisting largely of water, protein, antibodies, and vitamins. The production of milk continues practically as long as the baby continues to suckle.

lacteal
Small vessel responsible for absorbing fat in the small intestine. Occurring in the fingerlike villi of the ileum, lacteals have a milky appearance and drain into the lymphatic system. Before fat can pass into the lacteal, bile from the liver causes its emulsification into droplets small enough for attack by the enzyme lipase. The products of this digestion form into even smaller droplets, which diffuse into the villi. Large droplets reform before entering the lacteal and this causes the milky appearance.

lactic acid, or 2-hydroxypropanoic acid
Organic acid ($CH_3CHOHCOOH$), a colorless, almost odorless liquid, produced by certain bacteria during fermentation and by active muscle cells when they are exercised hard and are experiencing oxygen debt. An accumulation of lactic acid in the muscles may cause cramp. It occurs in yogurt, buttermilk, sour cream, poor wine, and certain plant extracts and is used in food preservation and in the preparation of pharmaceuticals.

lactose
White sugar, found in solution in milk; it forms 5 percent of cow's milk. It is commercially prepared from the whey obtained in cheese making. Like table sugar (sucrose), it is a disaccharide, consisting of two basic sugar units (monosaccharides), in this case, glucose and galactose. Unlike sucrose, it is tasteless.

lamina
In flowering plants (angiosperms), the blade of the leaf on either side of the midrib. The lamina is generally thin and flattened and is usually the primary organ of photosynthesis. It has a network of veins through which water and nutrients are conducted. More generally, a lamina is any thin, flat plant structure, such as the thallus of many seaweeds.

lamp shell
See **brachiopod.**

larva
Stage between hatching and adulthood in those species in which the young have a different appearance and way of life from the adults. Examples include tadpoles

(frogs) and caterpillars (butterflies and moths). Larvae are typical of the invertebrates, some of which (for example, shrimps) have two or more distinct larval stages. Among vertebrates, it is only the amphibians and some fishes that have a larval stage.

larynx

In mammals, a cavity at the upper end of the trachea (windpipe) containing the vocal cords. It is stiffened with cartilage and lined with mucous membrane. Amphibians and reptiles have much simpler larynxes, with no vocal cords. Birds have a similar cavity, called the syrinx, found lower down the trachea, where it branches to form the bronchi. It is very complex, with well-developed vocal cords.

lateral line system

System of sense organs in fishes and larval amphibians (tadpoles) that detects water movement. It usually consists of a row of interconnected pores on either side of the body that divides into a system of canals across the head.

leaf

Lateral outgrowth on the stem of a plant, and in most species the primary organ of photosynthesis. The chief leaf types are cotyledons (seed leaves), scale leaves (on underground stems), foliage leaves, and bracts (in the axil of which a flower is produced).

lecithin

Lipid (fat), containing nitrogen and phosphorus, that forms a vital part of the cell membranes of plant and animal cells. The name is from the Greek *lekithos,* meaning "egg yolk," eggs being a major source of lecithin.

legume

Plant of the family Leguminosae, which has a pod containing dry seeds. The family includes peas, beans, lentils, clover, and alfalfa (lucerne). Legumes are important in agriculture because of their specialized roots, which have nodules containing bacteria capable of fixing nitrogen from the air and increasing the fertility of the soil. The edible seeds of legumes are called pulses.

lek

A closely spaced set of very small territories each occupied by a single male during the mating season. Leks are found in the mating systems of several ground-dwelling birds (such as grouse) and a few antelopes.

lenticel

Small pore on the stems of woody plants or the trunks of trees. Lenticels are a means of gas exchange between the stem interior and the atmosphere. They consist of loosely packed cells with many air spaces in between and are easily seen on smooth-barked trees such as cherries, where they form horizontal lines on the trunk.

leucine

One of the nine essential amino acids.

leukocyte

See **white blood cell.**

life

The ability to grow, reproduce, and respond to such stimuli as light, heat, and sound. Life on earth may have began about four billion years ago when a chemical reaction produced the first organic substance. Over time, life has evolved from primitive single-celled organisms to complex multicellular ones. There are now some ten million different species of plants and animals living on the earth. The earliest fossil evidence of life is threadlike chains of cells discovered in 1980 in deposits in northwest Australia; these have been dated as being 3.5 billion years old.

life cycle

The sequence of developmental stages through which members of a given species pass. Most vertebrates have a simple life cycle consisting of fertilization of sex cells or gametes, a period of development as an embryo, a period of juvenile growth after hatching or birth, an adulthood including sexual reproduction, and finally death. Invertebrate life cycles are generally more complex and may involve major reconstitution of the individual's appearance (metamorphosis) and completely different styles of life. Plants have a special type of life cycle with two distinct phases, known as alternation of generations. Many insects such as cicadas, dragonflies, and mayflies have a long larvae or pupae phase and a short adult phase. Dragonflies live an aquatic life as larvae and an aerial life during the adult phase. In many invertebrates and protozoa there is a sequence of stages in the life cycle, and in parasites different stages often occur in different host organisms.

ligament

Strong, flexible connective tissue, made of the protein collagen, which joins bone to bone at moveable joints and sometimes encloses the joints. Ligaments prevent bone dislocation (under normal circumstances) but allow joint flexion. The ligaments around the joints are composed of white fibrous tissue. Other ligaments are composed of yellow elastic tissue, which is adapted to support a continuous but varying stress, as in the ligament connecting the various cartilages of the larynx (voice box).

lignin

Naturally occurring substance produced by plants to strengthen their tissues. It is difficult for enzymes to attack lignin, so living organisms cannot digest wood, with the exception of a few specialized fungi and bacteria. Lignin is the essential ingredient of all wood and is, therefore, of great commercial importance.

limiting factor

Any factor affecting the rate of a metabolic reaction. Levels of light or of carbon dioxide are limiting factors in photosynthesis because both are necessary for the production of carbohydrates. In experiments, photosynthesis is observed to slow down and eventually stop as the levels of light decrease.

linkage

In genetics, the association between two or more genes that tend to be inherited together because they are on the same chromosome. The closer together they are on the chromosome, the less likely they are to be separated by crossing over (one of the processes of recombination) and they are then described as being "tightly linked."

lipase

Enzyme responsible for breaking down fats into fatty acids and glycerol. It is produced by the pancreas and requires a slightly alkaline environment. The products of fat digestion are absorbed by the intestinal wall.

lipid

Any of a large number of esters of fatty acids, commonly formed by the reaction of a fatty acid with glycerol. They are soluble in alcohol but not in water. Lipids are the chief constituents of plant and animal waxes, fats, and oils.

liver

Large organ of vertebrates, which has many regulatory and storage functions. The human liver is situated in the upper abdomen and weighs about 2 kg/4.5 lb. It is divided into four lobes. The liver receives the products of digestion, converts glucose to glycogen (a long-chain carbohydrate used for storage), and then back to glucose when needed. In this way the liver regulates the level of glucose in the blood. It removes excess amino acids from the blood, converting them to urea, which is excreted by the kidneys. The liver also synthesizes vitamins, produces bile and blood-clotting factors, and removes damaged red cells and toxins such as alcohol from the blood.

liverwort

Nonvascular plant (with no "veins" to carry water and food), of the class Hepaticae, order Bryophyta, related to hornworts and mosses; it is found growing in damp places.

lomentum

Fruit similar to a pod but constricted between the seeds. When ripe, it splits into one-seeded units, as seen, for example, in the fruit of sainfoin *Onobrychis viciifolia* and radish *Raphanus raphanistrum*. It is a type of schizocarp.

lumen
The space enclosed by an organ, such as the bladder, or a tubular structure, such as the gastrointestinal tract.

lung
Large cavity of the body, used for gas exchange. It is essentially a sheet of thin, moist membrane that is folded so as to occupy less space. Most tetrapod (four-limbed) vertebrates have a pair of lungs occupying the thorax. The lung tissue, consisting of multitudes of air sacs and blood vessels, is very light and spongy, and functions by bringing inhaled air into close contact with the blood so that oxygen can pass into the organism and waste carbon dioxide can be passed out. The efficiency of lungs is enhanced by breathing movements, by the thinness and moistness of their surfaces, and by a constant supply of circulating blood.

luteinizing hormone
Hormone produced by the pituitary gland. In males, it stimulates the testes to produce androgens (male sex hormones). In females, it works together with follicle-stimulating hormone to initiate production of egg cells by the ovary. If fertilization occurs, it plays a part in maintaining the pregnancy by controlling the levels of the hormones estrogen and progesterone in the body.

lymph
Fluid found in the lymphatic system of vertebrates.

lymph nodes
Small masses of lymphatic tissue in the body that occur at various points along the major lymphatic vessels. Tonsils and adenoids are large lymph nodes. As the lymph passes through them it is filtered, and bacteria and other microorganisms are engulfed by cells known as macrophages.

lymphocyte
Type of white blood cell with a large nucleus, produced in the bone marrow. Most occur in the lymph and blood and around sites of infection. B lymphocytes or B cells are responsible for producing antibodies. T lymphocytes or T cells have several roles in the mechanism of immunity.

lymphokines
Chemical messengers produced by lymphocytes that carry messages among the cells of the immune system. Examples include interferon, which initiates defensive reactions to viruses, and the interleukins, which activate specific immune cells.

lysis
Any process that destroys a cell by rupturing its membrane or cell wall.

lysosome

Membrane-enclosed structure, or organelle, inside a cell, principally found in animal cells. Lysosomes contain enzymes that can break down proteins and other biological substances. They play a part in digestion, and in the white blood cells known as phagocytes the lysosome enzymes attack ingested bacteria.

macrophage

Type of white blood cell, or leukocyte, found in all vertebrate animals. Macrophages specialize in the removal of bacteria and other microorganisms, or of cell debris after injury. Like phagocytes, they engulf foreign matter, but they are larger than phagocytes and have a longer life span. They are found throughout the body, but mainly in the lymph and connective tissues, and especially the lungs, where they ingest dust, fibers, and other inhaled particles.

"magic bullet"

Term sometimes used for a drug that is specifically targeted on certain cells or tissues in the body, such as a small collection of cancerous cells or cells that have been invaded by a virus. Such drugs can be made in various ways, but monoclonal antibodies are increasingly being used to direct the drug to a specific target.

malaria

Infectious parasitic disease of the tropics transmitted by mosquitoes, marked by periodic fever and an enlarged spleen. When a female mosquito of the *Anopheles* genus bites a human who has malaria, it takes in with the human blood one of four malaria protozoa of the genus *Plasmodium*. This matures within the insect and is then transferred when the mosquito bites a new victim. Malaria affects about 267 million people in 103 countries, and in 1995 around 2.1 million people died of the disease. In sub-Saharan Africa alone between 1.5 and 2 million children die from malaria and its consequences each year.

maltase

Enzyme found in plants and animals that breaks down the disaccharide maltose into glucose.

maltose

A disaccharide sugar ($C_{12}H_{22}O_{11}$) in which both monosaccharide units are glucose.

mammal

Any of a large group of warm-blooded vertebrate animals of the class Mammalia characterized by having mammary glands in the female; these are used for suckling the young. Other features of mammals are hair (very reduced in some species, such as whales); a middle ear formed of three small bones (ossicles); a lower jaw consisting of two bones only; seven vertebrae in the neck; and no nucleus in the red blood cells. Mammals are divided into three groups:

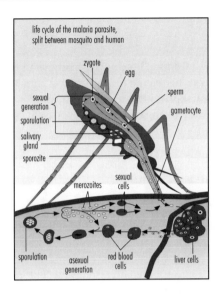

life cycle of the malaria parasite, split between mosquito and human

zygote

egg

sexual generation

sperm

sporulation

gametocyte

salivary gland

sporozite

merozoites

sexual cells

sporulation

asexual generation

red blood cells

liver cells

malaria

placental mammals, whose young develop inside the mother's body, in the uterus, receiving nourishment from the blood of the mother via the placenta;

marsupials, whose young are born at an early stage of development and develop further in a pouch on the mother's body where they are attached to and fed from a nipple; and

monotremes, where the young hatch from an egg outside the mother's body and are then nourished with milk.

The monotremes are the least evolved and have been largely displaced by more sophisticated marsupials and placentals, so that there are only a few types surviving (platypus and echidna). Placentals have spread to all parts of the globe, and where placentals have competed with marsupials, the placentals have in general displaced marsupial types. However, marsupials occupy many specialized niches in South America and, especially, Australasia. According to the *Red List of Threatened Animals* published by the International Union for the Conservation of Nature (IUCN) for 1996, 25 percent of mammal species are threatened with extinction.

mammary gland

In female mammals, a milk-producing gland derived from epithelial cells underlying the skin, active only after the production of young. In all but monotremes (egg-laying mammals), the mammary glands terminate in teats, which aid infant suckling. The number of glands and their position vary between species. In humans there are two, in cows four, and in pigs between ten and fourteen.

marsupial

From Greek *marsupion,* meaning "little purse," mammal in which the female has a pouch where she carries her young (born tiny and immature) for a considerable time after birth. Marsupials include omnivorous, herbivorous, and carnivorous species, among them the kangaroo, wombat, opossum, phalanger, bandicoot, dasyure, and wallaby.

matrix

Usually refers to the extracellular matrix.

medulla

Central part of an organ. In the mammalian kidney, the medulla lies beneath the outer cortex and is responsible for the reabsorption of water from the filtrate. In plants, it is a region of packing tissue in the center of the stem. In the vertebrate brain, the medulla is the posterior region responsible for the coordination of basic activities, such as breathing and temperature control.

medusa

The free-swimming phase in the life cycle of a coelenterate, such as a jellyfish or coral. The other phase is the sedentary polyp.

meiosis

A process of cell division in which the number of chromosomes in the cell is halved. It only occurs in eukaryotic cells and is part of a life cycle that involves sexual reproduction because it allows the genes of two parents to be combined without the total number of chromosomes increasing.

melanin

Brown pigment that gives color to the eyes, skin, hair, feathers, and scales of many vertebrates. In humans, melanin helps protect the skin against ultraviolet radiation from sunlight. Both genetic and environmental factors determine the amount of melanin in the skin.

melanism

Black coloration of animal bodies caused by large amounts of the pigment melanin. Melanin is of significance in insects, because melanic ones warm more rapidly in sunshine than do pale ones and can be more active in cool weather. A fall in temperature may stimulate such insects to produce more melanin. In industrial areas, dark insects and pigeons match sooty backgrounds and escape predation, but they are at a disadvantage in rural areas where they do not match their backgrounds. This is known as *industrial melanism.*

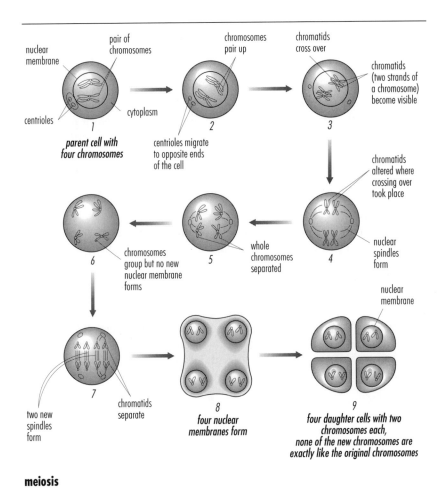

nuclear membrane

pair of chromosomes

chromosomes pair up

chromatids cross over

chromatids (two strands of a chromosome) become visible

cytoplasm

centrioles

1

parent cell with four chromosomes

centrioles migrate to opposite ends of the cell

2

3

chromatids altered where crossing over took place

nuclear spindles form

whole chromosomes separated

4

chromosomes group but no new nuclear membrane forms

6

5

nuclear membrane

two new spindles form

chromatids separate

7

8

four nuclear membranes form

9

four daughter cells with two chromosomes each, none of the new chromosomes are exactly like the original chromosomes

meiosis

membrane

In living things, a continuous layer, made up principally of fat molecules, that encloses a cell or organelles within a cell. Small molecules, such as water and sugars, can pass through the cell membrane by diffusion. Large molecules, such as proteins, are transported across the membrane via special channels, a process often involving energy input. The Golgi apparatus within the cell is thought to produce certain membranes.

Mendelism

In genetics, the theory of inheritance originally outlined by Austrian biologist Gregor Mendel in the mid-1800s. He suggested that, in sexually reproducing species, all characteristics are inherited through indivisible "factors" (now identified with genes) contributed by each parent to its offspring.

meniscus
The fibro-cartilage in joints, such as the knee joint.

menopause
In women, the cessation of reproductive ability, characterized by menstruation becoming irregular and eventually ceasing. The onset is at about the age of fifty, but varies greatly. Menopause is usually uneventful, but some women suffer from complications such as flushing, excessive bleeding, and nervous disorders. Since the 1950s, hormone-replacement therapy (HRT), using estrogen alone or with progestogen, a synthetic form of progesterone, has been developed to counteract such effects.

menstrual cycle
Cycle that occurs in female mammals of reproductive age, in which the body is prepared for pregnancy. At the beginning of the cycle, a Graafian (egg) follicle develops in the ovary, and the inner wall of the uterus forms a soft spongy lining. The egg is released from the ovary, and the uterine lining (endometrium) becomes vascularized (filled with blood vessels). If fertilization does not occur, the corpus luteum (remains of the Graafian follicle) degenerates, and the uterine lining breaks down and is shed. This is what causes the loss of blood that marks menstruation. The cycle then begins again. Human menstruation takes place from puberty to menopause, except during pregnancy, occurring about every 28 days.

meristem
Region of plant tissue containing cells that are actively dividing to produce new tissues (or have the potential to do so). Meristems found in the tip of roots and stems, the apical meristems, are responsible for the growth in length of these organs.

mesoglea
Layer of jellylike noncellular tissue that separates the endoderm and ectoderm in jellyfish and other coelenterates.

mesophyll
The tissue between the upper and lower epidermis of a leaf blade (lamina), consisting of parenchyma-like cells containing numerous chloroplasts.

metabolism
The chemical processes of living organisms enabling them to grow and to function. It involves a constant alternation of building up complex molecules (anabolism) and breaking them down (catabolism). For example, green plants build up complex organic substances from water, carbon dioxide, and mineral salts (photosynthesis); by digestion animals partially break down complex organic substances, ingested as food, and subsequently resynthesize them for use in their

own bodies. Within cells, complex molecules are broken down by the process of respiration. The waste products of metabolism are removed by excretion.

metamorphosis

Period during the life cycle of many invertebrates, most amphibians, and some fish, during which the individual's body changes from one form to another through a major reconstitution of its tissues. For example, adult frogs are produced by metamorphosis from tadpoles, and butterflies are produced from caterpillars following metamorphosis within a pupa. In classical thought and literature, metamorphosis is the transformation of a living being into another shape, either living or inanimate (as, for example, in the myth of Niobe, who, while weeping for her lost children, turned to stone). The Roman poet Ovid wrote about this theme.

metazoa

Another name for animals. It reflects an earlier system of classification, in which there were two main divisions within the animal kingdom, the multicellular animals, or metazoa, and the single-celled "animals" or protozoa. The protozoa are no longer included in the animal kingdom, so only the metazoa remain. *See also* **animals.**

methanogenic bacteria

One of a group of primitive microorganisms, the Archaea. They give off methane gas as a by-product of their metabolism, and are common in sewage treatment plants and hot springs, where the temperature is high and oxygen is absent. Archeons were originally classified as bacteria, but were found to be unique in 1996 following the gene sequencing of the deep-sea vent *Methanococcus jannaschii.*

methionine

One of the nine essential amino acids. It is also used as an antidote to paracetamol poisoning.

microbe

See **microorganism.**

microbiology

The study of microorganisms, mostly viruses and single-celled organisms such as bacteria, protozoa, and yeasts. The practical applications of microbiology are in medicine (since many microorganisms cause disease); in brewing, baking, and other food and beverage processes, where the microorganisms carry out fermentation; and in genetic engineering, which is creating increasing interest in the field of microbiology.

microorganism, or microbe

Living organism invisible to the naked eye but visible under a microscope. Microorganisms include viruses and single-celled organisms such as bacteria, protozoa, yeasts, and some algae. The term has no taxonomic significance in biology. The study of microorganisms is known as microbiology.

micropropagation

The mass production of plants by placing tiny pieces of plant tissue in sterile glass containers along with nutrients. Perfect clones of superplants are produced in sterile cabinets, with filtered air and carefully controlled light, temperature, and humidity. The system is used for the house-plant industry and for forestry—micropropagation gives immediate results, whereas obtaining genetically homogenous tree seed by traditional means would take more than one hundred years.

micropyle

In flowering plants, a small hole toward one end of the ovule. At pollination the pollen tube growing down from the stigma eventually passes through this pore. The male gamete is contained within the tube and is able to travel to the egg in the interior of the ovule. Fertilization can then take place, with subsequent seed formation and dispersal.

microscope

Instrument for forming magnified images with high resolution for detail. Optical and electron microscopes are the ones chiefly in use; other types include acoustic, scanning tunneling, and atomic force microscopes.

microtubules

Tiny tubes found in almost all cells with a nucleus. They help to define the shape of a cell by forming scaffolding for cilia and they also form the fibers of mitotic spindle.

migration

The movement, either seasonal or as part of a single life cycle, of certain animals, chiefly birds and fish, to distant breeding or feeding grounds.

milk

Secretion of the mammary glands of female mammals, with which they suckle their young during lactation. Over 85 percent is water, the remainder comprising protein, fat, lactose (a sugar), calcium, phosphorus, iron, and vitamins. The milk of cows, goats, and sheep is often consumed by humans, but regular drinking of milk after infancy is principally a Western practice.

milk teeth, or deciduous teeth

Teeth that erupt in childhood between the ages of six and thirty months. They are replaced by the permanent teeth, which erupt between the ages of six and twenty-one years.

mimicry

Imitation of one species (or group of species) by another. The most common form is *Batesian mimicry* (named for English naturalist H. W. Bates), where the mimic resembles a model that is poisonous or unpleasant to eat, and has aposematic, or warning, coloration; the mimic thus benefits from the fact that predators have learned to avoid the model. Hoverflies that resemble bees or wasps are an example. Appearance is usually the basis for mimicry, but calls, songs, scents, and other signals can also be mimicked.

hoverfly   wasp

harmless hoverfly mimics unpleasant wasp

mineral salt

In nutrition, a simple inorganic chemical that is required by living organisms. Plants usually obtain their mineral salts from the soil, while animals get theirs from their food. Important mineral salts include iron salts (needed by both plants and animals), magnesium salts (needed mainly by plants, to make chlorophyll), and calcium salts (needed by animals to make bone or shell). A trace element is required only in tiny amounts.

miscarriage

Spontaneous expulsion of a fetus from the womb before it is capable of independent survival. Often, miscarriages are due to an abnormality in the developing fetus.

mitochondria

Singular *mitochondrion,* membrane-enclosed organelles within eukaryotic cells, containing enzymes responsible for energy production during aerobic respiration. Mitochondria absorb O_2 and glucose and produce energy in the form of ATP by breaking down the glucose to CO_2 and H_2O. These rodlike or spherical bodies are thought to be derived from free-living bacteria that, at a very early stage in the history of life, invaded larger cells and took up a symbiotic way of life inside. Each still contains its own small loop of DNA called mitochondrial DNA, and new mitochondria arise by division of existing ones.

mitosis

The process of cell division by which identical daughter cells are produced. During mitosis the DNA is duplicated and the chromosome number doubled, so new cells contain the same amount of DNA as the original cell. The genetic material of eukaryotic cells is carried on a number of chromosomes. To control movements of chromosomes during cell division so that both new cells get the correct number, a system of protein tubules, known as the spindle, organizes the chromosomes into position in the middle of the cell before they replicate. The spindle then controls the movement of chromosomes as the cell goes through the stages of division: interphase, prophase, metaphase, anaphase, and telophase.

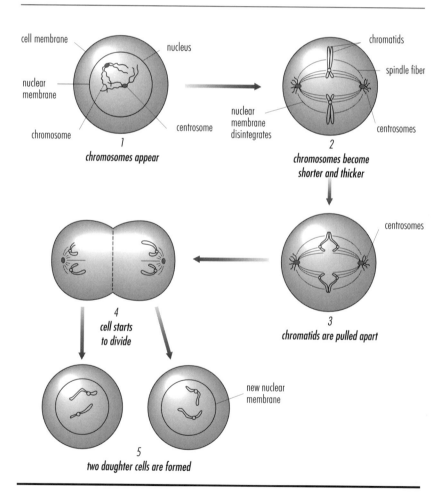

mitral valve

See bicuspid valve.

molar

One of the large teeth found toward the back of the mammalian mouth. The structure of the jaw, and the relation of the muscles, allows a massive force to be applied to molars. In herbivores the molars are flat with sharp ridges of enamel and are used for grinding, an adaptation to a diet of tough plant material. Carnivores have sharp powerful molars called carnassials, which are adapted for cutting meat.

mold

Furlike growth caused by any of a group of fungi living on foodstuffs and other organic matter; a few are parasitic on plants, animals, or each other. Many molds are of medical or industrial importance; for example, the antibiotic penicillin comes from a type of mold.

molecular biology

Study of the molecular basis of life, including the biochemistry of molecules such as DNA, RNA, and proteins, and the molecular structure and function of the various parts of living cells.

molecular clock

Use of rates of mutation in genetic material to calculate the length of time elapsed since two related species diverged from each other during evolution. The method can be based on comparisons of the DNA or of widely occurring proteins, such as hemoglobin.

mollusk

Any of a group of invertebrate animals of the phylum Molluska, most of which have a body divided into three parts: a head, a central mass containing the main organs, and a foot for movement; the more sophisticated octopuses and related mollusks have arms to capture their prey. The majority of mollusks are marine animals, but some live in fresh water, and a few live on dry land. They include clams, mussels, and oysters (bivalves), snails and slugs (gastropods), and cuttlefish, squids, and octopuses (cephalopods). The body is soft, without limbs (except for the cephalopods), and coldblooded. There is no internal skeleton, but many species have a hard shell covering the body.

molting

Periodic shedding of the hair or fur of mammals, feathers of birds, or skin of reptiles. In mammals and birds, molting is usually seasonal and is triggered by changes of day length.

monocarpic, or hapaxanthic

Describing plants that flower and produce fruit only once during their life cycle, after which they die. Most annual plants and biennial plants are monocarpic, but there are also a small number of monocarpic perennial plants that flower just once, sometimes after as long as ninety years, dying shortly afterwards, for example, century plant *Agave* and some species of bamboo *Bambusa*. The general biological term related to organisms that reproduce only once during their lifetime is semelparity.

monocotyledon

Angiosperm (flowering plant) having an embryo with a single cotyledon, or seed leaf (as opposed to dicotyledons, which have two). Monocotyledons usually have narrow leaves with parallel veins and smooth edges and hollow or soft stems. Their flower parts are arranged in threes. Most are small plants such as orchids, grasses, and lilies, but some are trees such as palms.

monocyte

Type of white blood cell. They are found in the tissues of the lymphatic and circulatory systems where their purpose is to remove foreign particles, such as bacteria and tissue debris, by ingesting them.

monoecious

Having separate male and female flowers on the same plant. Corn *(Zea mays)*, for example, has a tassel of male flowers at the top of the stalk and a group of female flowers (on the ear, or cob) lower down. Monoecism is a way of avoiding self-fertilization. Dioecious plants have male and female flowers on separate plants.

monohybrid inheritance

Pattern of inheritance seen in simple genetics experiments, where the two animals (or two plants) being crossed are genetically identical except for one gene.

monosaccharide, or simple sugar

Carbohydrate that cannot be hydrolysed (split) into smaller carbohydrate units. Examples are glucose and fructose, both of which have the molecular formula $C_6H_{12}O_6$.

monotreme

Any of a small group of primitive egg-laying mammals of the order Monotremata, found in Australasia. They include the echidnas (spiny anteaters) and the platypus.

morphology

The study of the physical structure and form of organisms, in particular their soft tissues.

moss

Small nonflowering plant of the class Musci (ten thousand species), forming with the liverworts and the hornworts the order Bryophyta. The stem of each plant bears rhizoids that anchor it; there are no true roots. Leaves spirally arranged on its lower portion have sexual organs at their tips. Most mosses flourish best in damp conditions where other vegetation is thin.

motility

The ability to move spontaneously. The term is often restricted to those cells that are capable of independent locomotion, such as spermatozoa. Many single-celled organisms are motile, for example, the ameba. Research has shown that cells capable of movement, including vertebrate muscle cells, have certain biochemical features in common. Filaments of the proteins actin and myosin are associated with motility, as are the metabolic processes needed for breaking down the energy-rich compound adenosine triphosphate (ATP).

motor nerve

In anatomy, any nerve that transmits impulses from the central nervous system to muscles or organs. Motor nerves cause voluntary and involuntary muscle contractions and stimulate glands to secrete hormones.

mouth

Cavity forming the entrance to the digestive tract. In land vertebrates, air from the nostrils enters the mouth cavity to pass down the trachea. The mouth in mammals is enclosed by the jaws, cheeks, and palate.

mucous membrane

Thin skin lining all animal body cavities and canals that come into contact with the air (for example, eyelids, breathing and digestive passages, genital tract). It secretes mucus, a moistening, lubricating, and protective fluid.

mucus

Lubricating and protective fluid, secreted by mucous membranes in many different parts of the body. In the gut, mucus smooths the passage of food and keeps potentially damaging digestive enzymes away from the gut lining. In the lungs, it traps airborne particles so that they can be expelled.

multiple birth

In humans, the production of more than two babies from one pregnancy. Multiple births can be caused by more than two eggs being produced and fertilized (often as the result of hormone therapy to assist pregnancy), or by a single fertilized egg dividing more than once before implantation.

muscle

Contractile animal tissue that produces locomotion and power and maintains the movement of body substances. Muscle is made of long cells that can contract to between one-half and one-third of their relaxed length.

mushroom
Fruiting body of certain fungi, consisting of an upright stem and a spore-producing cap with radiating gills on the undersurface. There are many edible species belonging to the genus *Agaricus,* including the field mushroom *A. campestris.*

mutagen
Any substance that increases the rate of gene mutation. A mutagen may also act as a carcinogen.

mutation
A change in the genes produced by a change in the DNA that makes up the hereditary material of all living organisms. Mutations, the raw material of evolution, result from mistakes during replication (copying) of DNA molecules. Only a few improve the organism's performance and are therefore favored by natural selection. Mutation rates are increased by certain chemicals and by radiation.

mutualism
See symbiosis.

mycelium
Interwoven mass of threadlike filaments, or *hyphae,* forming the main body of most fungi. The reproductive structures, or "fruiting bodies," grow from the mycelium.

mycorrhiza
Mutually beneficial (mutualistic) association occurring between plant roots and a soil fungus. Mycorrhizal roots take up nutrients more efficiently than nonmycorrhizal roots, and the fungus benefits by obtaining carbohydrates from the plant or tree.

myelin sheath
Insulating layer that surrounds nerve cells in vertebrate animals. It serves to speed up the passage of nerve impulses. Myelin is made up of fats and proteins and is formed from up to a hundred layers, laid down by special cells, the Schwann cells.

myoglobin
Globular protein, closely related to hemoglobin and located in vertebrate muscle. Oxygen binds to myoglobin and is released only when the hemoglobin can no longer supply adequate oxygen to muscle cells.

myrmecophyte
Plant that lives in association with a colony of ants and possesses specialized organs in which the ants live. For example, *Myrmecodia,* an epiphytic plant

from Malaysia, develops root tubers containing a network of cavities inhabited by ants.

nail
A hard, flat, flexible outgrowth of the digits of primates (humans, monkeys, and apes). Nails are composed of keratin.

nastic movement
Plant movement that is caused by an external stimulus, such as light or temperature, but is directionally independent of its source, unlike tropisms. Nastic movements occur as a result of changes in water pressure within specialized cells or differing rates of growth in parts of the plant. Examples include the opening and closing of crocus flowers following an increase or decrease in temperature (thermonasty), and the opening and closing of evening-primrose *Oenothera* flowers on exposure to dark and light (photonasty).

natural selection
The process whereby gene frequencies in a population change through certain individuals' producing more descendants than others because they are better able to survive and reproduce in their environment. The accumulated effect of natural selection is to produce adaptations such as the insulating coat of a polar bear or the spadelike forelimbs of a mole. The process is slow, relying firstly on random variation in the genes of an organism being produced by mutation and secondly on the genetic recombination of sexual reproduction. It was recognized by English naturalists Charles Darwin and Alfred Russel Wallace as the main process driving evolution.

nature-nurture controversy, or environment-heredity controversy
Long-standing dispute among philosophers and psychologists over the relative importance of environment; upbringing, experience, and learning ("nurture"), and heredity, that is, genetic inheritance ("nature"), in determining the makeup of an organism, as related to human personality and intelligence.

navel, or umbilicus
Small indentation in the center of the abdomen of mammals, marking the site of attachment of the umbilical cord, which connects the fetus to the placenta.

navigation, biological
The ability of animals or insects to navigate. Although many animals navigate by following established routes or known landmarks, many animals can navigate without such aids; for example, birds can fly several thousand miles back to their nest site, over unknown terrain. Such feats may be based on compass information derived from the position of the sun, moon, or stars, or on the characteristic patterns of earth's magnetic field.

neck

Structure between the head and the trunk in animals. In the back of the neck are the upper seven vertebrae of the spinal column, and there are many powerful muscles that support and move the head. In front, the neck region contains the pharynx and trachea, and behind these the esophagus. The large arteries (carotid, temporal, maxillary) and veins (jugular) that supply the brain and head are also located in the neck. The larynx (voice box) occupies a position where the trachea connects with the pharynx, and one of its cartilages produces the projection known as the Adam's apple. The thyroid gland lies just below the larynx and in front of the upper part of the trachea.

nectar

Sugary liquid secreted by some plants from a nectary, a specialized gland usually situated near the base of the flower. Nectar often accumulates in special pouches or spurs, not always in the same location as the nectary. Nectar attracts insects, birds, bats, and other animals to the flower for pollination and is the raw material used by bees in the production of honey.

nematode

Any of a group of unsegmented worms (phylum Nematoda) that are pointed at both ends, with a tough, smooth outer skin. They include many free-living species found in soil and water, including the sea, but a large number are parasites, such as the roundworms and pinworms that live in humans, or the eelworms that attack plant roots. They differ from flatworms in that they have two openings to the gut (a mouth and an anus).

neo-Darwinism

The modern theory of evolution, built up since the 1930s by integrating the nineteenth-century English scientist Charles Darwin's theory of evolution through natural selection with the theory of genetic inheritance founded in the work of the Austrian biologist Gregor Mendel.

neoteny

The retention of some juvenile characteristics in an animal that seems otherwise mature. An example is provided by the axolotl, a salamander that can reproduce sexually although still in its larval form.

nephron

Microscopic unit in vertebrate kidneys that forms urine. A human kidney is composed of over a million nephrons. Each nephron consists of a knot of blood capillaries called a glomerulus, contained in the Bowman's capsule, and a long narrow tubule enmeshed with yet more capillaries. Waste materials and water pass from the bloodstream into the tubule, and essential minerals and some water are reabsorbed from the tubule back into the bloodstream. The remaining filtrate (urine) is passed out from the body.

nerve

Bundle of nerve cells enclosed in a sheath of connective tissue and transmitting nerve impulses to and from the brain and spinal cord. A single nerve may contain both motor and sensory nerve cells, but they function independently.

nerve cell, or neuron

Elongated cell, the basic functional unit of the nervous system that transmits information rapidly between different parts of the body. Each nerve cell has a cell body, containing the nucleus, from which trail processes called dendrites, responsible for receiving incoming signals. The unit of information is the nerve impulse, a traveling wave of chemical and electrical changes involving the membrane of the nerve cell. The cell's longest process, the axon, carries impulses away from the cell body.

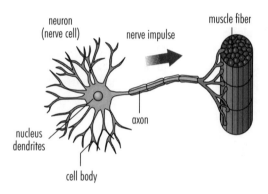

nervous system

The system of interconnected nerve cells of most invertebrates and all vertebrates. It is composed of the central and autonomic nervous systems. It may be as simple as the nerve net of coelenterates (for example, jellyfishes) or as complex as the mammalian nervous system, with a central nervous system comprising brain and spinal cord and a peripheral nervous system connecting up with sensory organs, muscles, and glands.

nest

Place chosen or constructed by a bird or other animal for incubation of eggs, hibernation, and shelter. Nests vary enormously, from saucerlike hollows in the ground, such as the scrapes of hares, to large and elaborate structures, such as the 4-m/13-ft-diameter mounds of the megapode birds.

neuron

See nerve cell.

neurotransmitter
Chemical that diffuses across a synapse, and thus transmits impulses between nerve cells, or between nerve cells and effector organs (for example, muscles). Common neurotransmitters are noradrenaline (which also acts as a hormone) and acetylcholine, the latter being most frequent at junctions between nerve and muscle. Nearly fifty different neurotransmitters have been identified.

niacin
See nicotinic acid.

niche
In ecology, the "place" occupied by a species in its habitat, including all chemical, physical, and biological components, such as what it eats, the time of day at which the species feeds, temperature, moisture, the parts of the habitat that it uses (for example, trees or open grassland), the way it reproduces, and how it behaves.

nicotinic acid, or niacin
Water-soluble vitamin ($C_5H_5N.COOH$) of the B complex, found in meat, fish, and cereals; it can also be formed in small amounts in the body from the essential amino acid tryptophan. Absence of nicotinic acid from the diet leads to the disease pellagra.

nitrate
Salt or ester of nitric acid, containing the NO_3^- ion. Nitrates are used in explosives, in the chemical and pharmaceutical industries, in curing meat, and as fertilizers. They are the most water-soluble salts known and play a major part in the nitrogen cycle. Nitrates in the soil, whether naturally occurring or from inorganic or organic fertilizers, can be used by plants to make proteins and nucleic acids. However, runoff from fields can result in nitrate pollution.

nitric oxide, or nitrogen monoxide
Colorless gas (NO) released when metallic copper reacts with nitric acid and when nitrogen and oxygen combine at high temperatures. It is oxidized to nitrogen dioxide on contact with air. Nitric oxide has a wide range of functions in the body. It is involved in the transmission of nerve impulses and the protection of nerve cells against stress. It is released by macrophages in the immune system in response to viral and bacterial infection or to the proliferation of cancer cells. It is also important in the control of blood pressure.

nitrification
Process that takes place in soil when bacteria oxidize ammonia, turning it into nitrates. Nitrates can be absorbed by the roots of plants, so this is a vital stage in the nitrogen cycle.

nitrogen cycle

The process of nitrogen passing through the ecosystem. Nitrogen, in the form of inorganic compounds (such as nitrates) in the soil, is absorbed by plants and turned into organic compounds (such as proteins) in plant tissue. A proportion of this nitrogen is eaten by herbivores, with some of this in turn being passed on to the carnivores, which feed on the herbivores. The nitrogen is ultimately returned to the soil as excrement and when organisms die and are converted back to inorganic form by decomposers.

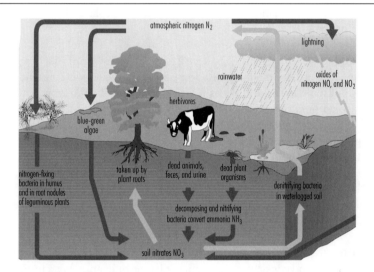

nitrogen fixation

The process by which nitrogen in the atmosphere is converted into nitrogenous compounds by the action of microorganisms, such as cyanobacteria and bacteria, in conjunction with certain legumes. Several chemical processes duplicate nitrogen fixation to produce fertilizers.

nitrogen monoxide

See nitric oxide.

nose

In humans, the upper entrance of the respiratory tract; the organ of the sense of smell. The external part is divided down the middle by a septum of cartilage. The nostrils contain plates of cartilage that can be moved by muscles and have a growth of stiff hairs at the margin to prevent foreign objects from entering. The whole nasal cavity is lined with a mucous membrane that warms and moistens the air as it enters and ejects dirt. In the upper parts of the cavity the membrane contains fifty million olfactory receptor cells (cells sensitive to smell).

nostril

In vertebrates, the opening of the nasal cavity, in which cells sensitive to smell are located. (In fish, these cells detect water-borne chemicals, so they are effectively organs of taste.) In vertebrates with lungs, the nostrils also take in air. In humans, and most other mammals, the nostrils are located on the nose.

notochord

The stiff but flexible rod that lies between the gut and the nerve cord of all embryonic and larval chordates, including the vertebrates. It forms the supporting structure of the adult lancelet, but in vertebrates it is replaced by the vertebral column, or spine.

nucleic acid

Complex organic acid made up of a long chain of nucleotides, present in the nucleus and sometimes the cytoplasm of the living cell. The two types, known as DNA (deoxyribonucleic acid) and RNA (ribonucleic acid), form the basis of heredity. The nucleotides are made up of a sugar (deoxyribose or ribose), a phosphate group, and one of four purine or pyrimidine bases. The order of the bases along the nucleic acid strand contains the genetic code.

nucleolus

A structure found in the nucleus of eukaryotic cells. It produces the RNA that makes up the ribosomes, from instructions in the DNA.

nucleotide

Organic compound consisting of a purine (adenine or guanine) or a pyrimidine (thymine, uracil, or cytosine) base linked to a sugar (deoxyribose or ribose) and a phosphate group. DNA and RNA are made up of long chains of nucleotides.

nucleus

The central, membrane-enclosed part of a eukaryotic cell, containing threads of DNA. During cell division these coil up to form chromosomes. The nucleus controls the function of the cell by determining which proteins are produced within it. Because proteins are the chief structural molecules of living matter and, as enzymes, regulate all aspects of metabolism, it may be seen that the genetic code within the nucleus is effectively responsible for building and controlling the whole organism.

nut

Any dry, single-seeded fruit that does not split open to release the seed, such as the chestnut. A nut is formed from more than one carpel, but only one seed becomes fully formed, the remainder aborting. The wall of the fruit, the pericarp, becomes hard and woody, forming the outer shell.

nutation

In botany, the spiral movement exhibited by the tips of certain stems during growth; it enables a climbing plant to find a suitable support. Nutation sometimes also occurs in tendrils and flower stalks.

nymph

In entomology, the immature form of insects that do not have a pupal stage; for example, grasshoppers and dragonflies. Nymphs generally resemble the adult (unlike larvae), but do not have fully formed reproductive organs or wings.

oil

Flammable substance, usually insoluble in water, and composed chiefly of carbon and hydrogen. Oils may be solids (fats and waxes) or liquids. The three main types are: essential oils, obtained from plants; fixed oils, obtained from animals and plants; and mineral oils, obtained chiefly from the refining of petroleum.

omnivore

Animal that feeds on both plant and animal material. Omnivores have digestive adaptations intermediate between those of herbivores and carnivores, with relatively unspecialized digestive systems and gut microorganisms that can digest a variety of foodstuffs. Omnivores include humans, the chimpanzee, the cockroach, and the ant.

oncogene

Gene carried by a virus that induces a cell to divide abnormally, giving rise to a cancer. Oncogenes arise from mutations in genes (proto-oncogenes) found in all normal cells. They are usually also found in viruses that are capable of transforming normal cells to tumor cells. Such viruses are able to insert their oncogenes into the host cell's DNA, causing it to divide uncontrollably. More than one oncogene may be necessary to transform a cell in this way.

onco-mouse

Mouse that has a human oncogene (gene that can cause certain cancers) implanted into its cells by genetic engineering. Such mice are used to test anticancer treatments and were patented in the United States by Harvard University in 1988, thereby protecting its exclusive rights to produce the animal and profit from its research.

ontogeny

Process of development of a living organism, including the part of development that takes place after hatching or birth. The idea that "ontogeny recapitulates phylogeny" (the development of an organism goes through the same stages as its evolutionary history), proposed by the German scientist Ernst Heinrich Haeckel, is now discredited.

oocyte
In medicine, an immature ovum. Only a fraction of the oocytes produced in the ovary survive until puberty and not all of these undergo meiosis to become an ovum that can be fertilized by a sperm.

oosphere
Another name for the female gamete, or ovum, of certain plants such as algae.

operon
Group of genes that are found next to each other on a chromosome and are turned on and off as an integrated unit. They usually produce enzymes that control different steps in the same biochemical pathway. Operons were discovered in 1961 by the French biochemists François Jacob and Jacques Monod in bacteria.

opiate, endogenous
Naturally produced chemical in the body that has effects similar to morphine and other opiate drugs; a type of neurotransmitter. Examples include endorphins and encephalins.

optic nerve
Large nerve passing from the eye to the brain, carrying visual information. In mammals, it may contain up to a million nerve fibers, connecting the sensory cells of the retina to the optical centers in the brain. Embryologically, the optic nerve develops as an outgrowth of the brain.

order
In biological classification, a group of related families. For example, the horse, rhinoceros, and tapir families are grouped in the order Perissodactyla, the odd-toed ungulates, because they all have either one or three toes on each foot. The names of orders are not shown in italic (unlike genus and species names) and by convention they have the ending "-formes" in birds and fish; "-a" in mammals, amphibians, reptiles, and other animals; and "-ales" in fungi and plants. Related orders are grouped together in a class.

organ
Part of a living body that has a distinctive function or set of functions. Examples include the liver or brain in animals, or the leaf in plants. An organ is composed of a group of coordinated tissues. A group of organs working together to perform a function is called an organ system, for example, the digestive system comprises a number of organs including the stomach, the small intestine, the colon, the pancreas, and the liver.

organelle
Discrete and specialized structure in a living cell; organelles include mitochondria, chloroplasts, lysosomes, ribosomes, and the nucleus.

organic compound

A class of compounds that contain carbon. The original distinction between organic and inorganic compounds was based on the belief that the molecules of living systems were unique and could not be synthesized in the laboratory. Today it is routine to manufacture thousands of organic chemicals both in research and in the drug industry.

organizer

In embryology, a part of the embryo that causes changes to occur in another part, through induction, thus "organizing" development and differentiation.

ornithology

Study of birds. It covers scientific aspects relating to their structure and classification, habits, song, flight, and value to agriculture as destroyers of insect pests. Worldwide scientific banding (the fitting of coded rings to captured specimens) has resulted in accurate information on bird movements and distribution. There is an International Council for Bird Preservation with its headquarters at the Natural History Museum, London.

ornithophily

Pollination of flowers by birds. Ornithophilous flowers are typically brightly colored, often red or orange. They produce large quantities of thin, watery nectar and are scentless because most birds do not respond well to smell. They are found mostly in tropical areas, with hummingbirds being important pollinators in North and South America, and sunbirds in Africa and Asia.

osmoregulation

Process whereby the water content of living organisms is maintained at a constant level. If the water balance is disrupted, the concentration of salts will be too high or too low, and vital functions, such as nerve conduction, will be adversely affected.

osmosis

Movement of water through a selectively permeable membrane separating solutions of different concentrations. Water passes by diffusion from a *weak solution* (high water concentration) to a *strong solution* (low water concentration) until the two concentrations are equal. The selectively permeable membrane allows the diffusion of water but not of the solute (e.g., sugar molecules). Many cell membranes behave in this way, and osmosis is a vital mechanism in the transportation of fluids in living organisms—for example, in the transportation of water from soil (weak solution) into the roots of plants (stronger solution of cell sap).

ossification, or osteogenesis

Process whereby bone is formed in vertebrate animals by special cells (osteoblasts) that secrete layers of extracellular matrix on the surface of the existing cartilage. Conversion to bone occurs through the deposition of calcium phosphate crystals within the matrix.

osteology
Part of the science of anatomy, dealing with the structure, function, and development of bones.

ovary
In female animals, the organ that generates the ovum. In humans, the ovaries are two whitish rounded bodies about 25 mm/1 in by 35 mm/1.5 in, located in the lower abdomen to either side of the uterus. Every month, from puberty to the onset of the menopause, an ovum is released from the ovary. This is called ovulation, and forms part of the menstrual cycle. In botany, an ovary is the expanded basal portion of the carpel of flowering plants, containing one or more ovules. It is hollow with a thick wall to protect the ovules. Following fertilization of the ovum, it develops into the fruit wall or pericarp.

oviduct
See **Fallopian tube.**

oviparous
Method of animal reproduction in which eggs are laid by the female and develop outside her body, in contrast to ovoviviparous and viviparous. It is the most common form of reproduction.

ovoviviparous
Method of animal reproduction in which fertilized eggs develop within the female (unlike oviparous) and the embryo gains no nutritional substances from the female (unlike viviparous). It occurs in some invertebrates, fishes, and reptiles.

ovulation
In female animals, the process of releasing egg cells (ova) from the ovary. In mammals it occurs as part of the menstrual cycle.

ovule
Structure found in seed plants that develops into a seed after fertilization. It consists of an embryo sac containing the female gamete (ovum or egg cell), surrounded by nutritive tissue, the nucellus. Outside this there are one or two coverings that provide protection, developing into the testa, or seed coat, following fertilization.

ovum
Plural *ova,* female gamete (sex cell) before fertilization. In animals it is called an egg and is produced in the ovaries. In plants, where it is also known as an egg cell or oosphere, the ovum is produced in an ovule. The ovum is nonmotile. It must

be fertilized by a male gamete before it can develop further, except in cases of parthenogenesis.

oxygen
Colorless, odorless, tasteless, nonmetallic, gaseous element, symbol O, atomic number 8, atomic weight 15.9994. Oxygen is a by-product of photosynthesis and the basis for respiration in plants and animals.

oxygen debt
Physiological state produced by vigorous exercise, in which the lungs cannot supply all the oxygen that the muscles need.

painkiller
Agent for relieving pain. Types of painkiller include analgesics such as aspirin and aspirin substitutes, morphine, codeine, paracetamol, and synthetic versions of the natural inhibitors, the encephalins and endorphins, which which do not produce the side effects of the others.

palate
In mammals, the roof of the mouth. The bony front part is the hard palate, the muscular rear part the soft palate. Incomplete fusion of the two lateral halves of the palate (cleft palate) causes interference with speech.

palisade cell
Cylindrical cell lying immediately beneath the upper epidermis of a leaf. Palisade cells normally exist as one closely packed row and contain many chloroplasts. During the hours of daylight palisade cells are photosynthetic, using the energy of the sun to create carbohydrates from water and carbon dioxide.

pancreas
In vertebrates, an accessory gland of the digestive system located close to the duodenum. When stimulated by the hormone secretin, it releases enzymes into the duodenum that digest starches, proteins, and fats. In humans, it is about 18 cm/7 in long, and lies behind and below the stomach. It contains groups of cells called the **islets of Langerhans,** which secrete the hormones insulin and glucagon that regulate the blood sugar level.

pantothenic acid
Water-soluble vitamin ($C_9H_{17}NO_5$) of the B complex, found in a wide variety of foods. Its absence from the diet can lead to dermatitis, and it is known to be involved in the breakdown of fats and carbohydrates.

pappus
Plural *pappi,* in botany, a modified calyx comprising a ring of fine, silky hairs, or sometimes scales or small teeth, that persists after fertilization. Pappi are found in

members of the daisy family (Compositae) such as the dandelions *Taraxacum,* where they form a parachutelike structure that aids dispersal of the fruit.

parasite
Organism that lives on or in another organism (called the host) and depends on it for nutrition, often at the expense of the host's welfare. Parasites that live inside the host, such as liver flukes and tapeworms, are called endoparasites; those that live on the exterior, such as fleas and lice, are called ectoparasites.

parathyroid
One of a pair of small endocrine glands. Most tetrapod vertebrates, including humans, possess two such pairs, located behind the thyroid gland. They secrete parathyroid hormone, which regulates the amount of calcium in the blood.

parenchyma
Plant tissue composed of loosely packed, more or less spherical cells, with thin cellulose walls. Although parenchyma often has no specialized function, it is usually present in large amounts, forming a packing or ground tissue. It usually has many intercellular spaces.

parental care
The time and energy spent by a parent in order to rear its offspring to maturity. Among animals, it ranges from the simple provision of a food supply for the hatching young at the time the eggs are laid (for example, many wasps) to feeding and protection of the young after hatching or birth, as in birds and mammals. In the more social species, parental care may include the teaching of skills—for example, female cats teach their kittens to hunt.

parthenocarpy
In botany, the formation of fruits without seeds. This phenomenon, of no obvious benefit to the plant, occurs naturally in some plants, such as bananas. It can also be induced in some fruit crops, either by breeding or by applying certain plant hormones.

parthenogenesis
Development of an ovum (egg) without any genetic contribution from a male. Parthenogenesis is the normal means of reproduction in a few plants (for example, dandelions) and animals (for example, certain fish). Some sexually reproducing species, such as aphids, show parthenogenesis at some stage in their life cycle to accelerate reproduction to take advantage of good conditions.

patella
Flat bone (the "kneecap"), embedded in the knee tendon of birds and mammals, which protects the joint from injury.

pathogen

From Greek, meaning "disease producing," in medicine, any microorganism that causes disease. Most pathogens are parasites, and the diseases they cause are incidental to their search for food or shelter inside the host. Nonparasitic organisms, such as soil bacteria or those living in the human gut and feeding on waste foodstuffs, can also become pathogenic to a person whose immune system or liver is damaged. The larger parasites that can cause disease, such as nematode worms, are not usually described as pathogens.

pectoral

Relating to the upper area of the thorax associated with the muscles and bones used in moving the arms or forelimbs, in vertebrates. In birds, the *pectoralis major* is the very large muscle used to produce a powerful downbeat of the wing during flight.

pedicel

The stalk of an individual flower, which attaches it to the main floral axis, often developing in the axil of a bract.

pellagra

Chronic disease mostly seen in subtropical countries in which the staple food is corn. It is caused by deficiency of nicotinic acid (one of the B vitamins), which is contained in protein foods, beans and peas, and yeast. Symptoms include diarrhea, skin eruptions, and mental disturbances.

pelvis

In vertebrates, the lower area of the abdomen featuring the bones and muscles used to move the legs or hindlimbs. The pelvic girdle is a set of bones that allows movement of the legs in relation to the rest of the body and provides sites for the attachment of relevant muscles.

penis

Male reproductive organ containing the urethra, the channel through which urine and semen are voided. It transfers sperm to the female reproductive tract to fertilize the ovum. In mammals, the penis is made erect by vessels that fill with blood, and in most mammals (but not humans) is stiffened by a bone.

pentadactyl limb

Typical limb of the mammals, birds, reptiles, and amphibians. These vertebrates (animals with backbone) are all descended from primitive amphibians whose immediate ancestors were fleshy-finned fish. The limb which evolved in those amphibians had three parts: a "hand/foot" with five digits (fingers/toes), a lower limb containing two bones, and an upper limb containing one bone.

pepsin

Enzyme that breaks down proteins during digestion. It requires a strongly acidic environment and is present in the stomach.

peptide

Molecule comprising two or more amino acid molecules (not necessarily different) joined by peptide bonds, whereby the acid group of one acid is linked to the amino group of the other. The number of amino acid molecules in the peptide is indicated by referring to it as a di-, tri-, or polypeptide (two, three, or many amino acids).

perennating organ

In plants, that part of a biennial plant or herbaceous perennial that allows it to survive the winter; usually a root, tuber, rhizome, bulb, or corm.

perennial plant

Plant that lives for more than two years. Herbaceous perennials have aerial stems and leaves that die each autumn. They survive the winter by means of an underground storage (perennating) organ, such as a bulb or rhizome. Trees and shrubs or woody perennials have stems that persist above ground throughout the year, and may be either deciduous or evergreen.

perianth

In botany, a collective term for the outer whorls of the flower, which protect the reproductive parts during development. In most dicotyledons the perianth is composed of two distinct whorls, the calyx of sepals and the corolla of petals, whereas in many monocotyledons the sepals and petals are indistinguishable and the segments of the perianth are then known individually as tepals.

pericarp

Wall of a fruit. It encloses the seeds and is derived from the ovary wall. In fruits such as the acorn, the pericarp becomes dry and hard, forming a shell around the seed. In fleshy fruits the pericarp is typically made up of three distinct layers. The epicarp, or exocarp, forms the tough outer skin of the fruit, while the mesocarp is often fleshy and forms the middle layers. The innermost layer or endocarp, which surrounds the seeds, may be membranous or thick and hard, as in the drupe (stone) of cherries, plums, and apricots.

peristalsis

Wavelike contractions, produced by the contraction of smooth muscle, that pass along tubular organs, such as the intestines. The same term describes the wavelike motion of earthworms and other invertebrates, in which part of the body contracts as another part elongates.

peritoneum

Membrane lining the abdominal cavity and digestive organs of vertebrates. Peritonitis, inflammation within the peritoneum, can occur due to infection or other irritation. It is sometimes seen following a burst appendix and quickly proves fatal if not treated.

perspiration

Excretion of water and dissolved substances from the sweat glands of the skin of mammals. Perspiration has two main functions: body cooling by the evaporation of water from the skin surface, and excretion of waste products such as salts.

pest

Any insect, fungus, rodent, or other living organism that has a harmful effect on human beings, other than those that directly cause human diseases. Most pests damage crops or livestock, but the term also covers those that damage buildings, destroy food stores, and spread disease.

petal

Part of a flower whose function is to attract pollinators such as insects or birds. Petals are frequently large and brightly colored and may also be scented. Some have a nectary at the base and markings on the petal surface, known as honey guides, to direct pollinators to the source of the nectar. In wind-pollinated plants, however, the petals are usually small and insignificant, and sometimes absent altogether. Petals are derived from modified leaves, and are known collectively as a corolla.

petiole

In botany, the stalk attaching the leaf blade, or lamina, to the stem. Typically it is continuous with the midrib of the leaf and attached to the base of the lamina, but occasionally it is attached to the lower surface of the lamina, as in the nasturtium (a peltate leaf). Petioles that are flattened and leaflike are termed phyllodes. Leaves that lack a petiole are said to be sessile.

phage

Another name for a bacteriophage, a virus that attacks bacteria.

phagocyte

Type of white blood cell, or leukocyte, that can engulf a bacterium or other invading microorganism. Phagocytes, which differ in size and life span, are found in blood, lymph, and other body tissues. They also ingest foreign matter and dead tissue.

phanerogam

Obsolete term for a plant that bears flowers or cones and reproduces by means of seeds, that is an angiosperm and gymnosperm, or a seed plant. Plants such as mosses, fungi, and ferns were known as cryptogams.

pharynx
Muscular cavity behind the nose and mouth, extending downward from the base of the skull. Its walls are made of muscle strengthened with a fibrous layer and lined with mucous membrane. The internal nostrils lead backward into the pharynx, which continues downward into the esophagus and (through the epiglottis) into the windpipe. On each side, a Eustachian tube enters the pharynx from the middle ear cavity.

phenotype
In genetics, visible traits, those actually displayed by an organism. The phenotype is not a direct reflection of the genotype because some alleles are masked by the presence of other, dominant alleles. The phenotype is further modified by the effects of the environment (for example, poor nutrition stunts growth).

phenylanaline
One of the nine essential amino acids. Phenylketonuria is a rare genetic disease which results from the inability to metabolize the phenylalanine present in food.

pheromone
Chemical signal (such as an odor) that is emitted by one animal and affects the behavior of others. Pheromones are used by many animal species to attract mates.

phloem
Tissue found in vascular plants whose main function is to conduct sugars and other food materials from the leaves, where they are produced, to all other parts of the plant.

phospholipid
Any lipid consisting of a glycerol backbone, a phosphate group, and two long chains. Phospholipids are found everywhere in living systems as the basis for biological membranes.

photochemical reaction
Any chemical reaction in which light is produced or light initiates the reaction. Light can initiate reactions by exciting atoms or molecules and making them more reactive: the light energy becomes converted to chemical energy. Many photochemical reactions set up a chain reaction and produce free radicals.

photolysis
Chemical reaction that is driven by light or ultraviolet radiation. For example, the light reaction of photosynthesis (the process by which green plants manufacture carbohydrates from carbon dioxide and water) is a photolytic reaction.

photoperiodism
Biological mechanism that determines the timing of certain activities by responding to changes in day length. The flowering of many plants is initiated in this way.

Photoperiodism in plants is regulated by a light-sensitive pigment, phytochrome. The breeding seasons of many temperate-zone animals are also triggered by increasing or declining day length, as part of their biorhythms.

photosynthesis

Process by which green plants trap light energy from the sun. This energy is used to drive a series of chemical reactions which lead to the formation of carbohydrates. The carbohydrates occur in the form of simple sugar, or glucose, which provides the basic food for both plants and animals. For photosynthesis to occur, the plant must possess chlorophyll and must have a supply of carbon dioxide and water. Photosynthesis takes place inside chloroplasts which are found mainly in the leaf cells of plants. The by-product of photosynthesis, oxygen, is of great importance to all living organisms, and virtually all atmospheric oxygen has originated by photosynthesis.

phototropism

Movement of part of a plant toward or away from a source of light. Leaves are positively phototropic, detecting the source of light and orienting themselves to receive the maximum amount.

phyllotaxis

The arrangement of leaves on a plant stem. Leaves are nearly always arranged in a regular pattern and in the majority of plants they are inserted singly, either in a spiral arrangement up the stem, or on alternate sides. Other principal forms are opposite leaves, where two arise from the same node, and whorled, where three or more arise from the same node.

phylogeny

Historical sequence of changes that occurs in a given species during the course of its evolution. It was once erroneously associated with ontogeny (the process of development of a living organism).

phylum

Plural *phyla,* major grouping in biological classification. Mammals, birds, reptiles, amphibians, fishes, and tunicates belong to the phylum Chordata; the phylum Molluska consists of snails, slugs, mussels, clams, squid, and octopuses; the phylum Porifera contains sponges; and the phylum Echinodermata includes starfish, sea urchins, and sea cucumbers. In classifying plants (where the term "division" often takes the place of "phylum"), there are between four and nine phyla depending on the criteria used; all flowering plants belong to a single phylum, Angiospermata, and all conifers to another, Gymnospermata. Related phyla are grouped together in a kingdom; phyla are subdivided into classes.

physiology

Branch of biology that deals with the functioning of living organisms, as opposed to anatomy, which studies their structures.

phytomenadione

One form of vitamin K, a fat-soluble chemical found in green vegetables. It is involved in the production of prothrombin, which is essential in blood clotting. It is given to newborns to prevent potentially fatal brain hemorrhages.

pineal body, or pineal gland

A cone-shaped outgrowth of the vertebrate brain. In some lower vertebrates, it develops a rudimentary lens and retina, which show it to be derived from an eye, or pair of eyes, situated on the top of the head in ancestral vertebrates. In fishes that can change color to match their background, the pineal perceives the light level and controls the color change. In birds, the pineal detects changes in daylight and stimulates breeding behavior as spring approaches. Mammals also have a pineal gland, but it is located deeper within the brain. It secretes a hormone, melatonin, thought to influence rhythms of activity. In humans, it is a small piece of tissue attached by a stalk to the rear wall of the third ventricle of the brain.

pinna

In botany, the primary division of a pinnate leaf. In mammals, the pinna is the external part of the ear.

pinnate leaf

Leaf that is divided up into many small leaflets, arranged in rows along either side of a midrib, as in ash trees *(Fraxinus)*. It is a type of compound leaf. Each leaflet is known as a pinna, and where the pinnae are themselves divided, the secondary divisions are known as pinnules.

pistil

General term for the female part of a flower, either referring to one single carpel or a group of several fused carpels.

pituitary gland

Major endocrine gland of vertebrates, situated in the center of the brain. It is attached to the hypothalamus by a stalk. The pituitary consists of two lobes. The posterior lobe is an extension of the hypothalamus, and is in effect nervous tissue. It stores two hormones synthesized in the hypothalamus: ADH and oxytocin. The anterior lobe secretes six hormones, some of which control the activities of other glands (thyroid, gonads, and adrenal cortex); others are direct-acting hormones affecting milk secretion and controlling growth.

placenta

Organ that attaches the developing embryo or fetus to the uterus in placental mammals (mammals other than marsupials, platypuses, and echidnas). Composed of maternal and embryonic tissue, it links the blood supply of the embryo to the blood supply of the mother, allowing the exchange of oxygen, nutrients, and waste products. The two blood systems are not in direct contact, but are separated by thin membranes, with materials diffusing across from one system to the other. The placenta also produces hormones that maintain and regulate pregnancy. It is shed as part of the afterbirth.

plankton

Small, often microscopic, forms of plant and animal life that live in the upper layers of fresh and salt water, and are an important source of food for larger animals. Marine plankton is concentrated in areas where rising currents bring mineral salts to the surface.

plant

Organism that carries out photosynthesis, has cellulose cell walls and complex cells, and is immobile. A few parasitic plants have lost the ability to photosynthesize but are still considered to be plants. Plants are autotrophs, that is, they make carbohydrates from water and carbon dioxide, and are the primary producers in all food chains, so that all animal life is dependent on them. They play a vital part in the carbon cycle, removing carbon dioxide from the atmosphere and generating oxygen. The study of plants is known as botany.

plant classification

Taxonomy or classification of plants. Originally the plant kingdom included bacteria, diatoms, dinoflagellates, fungi, and slime molds, but these are not now thought of as plants. The groups that are always classified as plants are the bryophytes (mosses and liverworts), pteridophytes (ferns, horsetails, and club mosses), gymnosperms (conifers, yews, cycads, and ginkgos), and angiosperms (flowering plants). The angiosperms are split into monocotyledons (for example, orchids, grasses, lilies) and dicotyledons (for example, oak, buttercup, geranium, and daisy).

plant hormone

Substance produced by a plant that has a marked effect on its growth, flowering, leaf fall, fruit ripening, or some other process. Examples include auxin, gibberellin, ethene, and cytokinin.

plant propagation

Production of plants. Botanists and horticulturalists can use a wide variety of means for propagating plants. There are the natural techniques of vegetative reproduction, together with cuttings, grafting, and micropropagation. The range is wide because most plant tissue, unlike animal tissue, can give rise to the complete range of tissue types within a particular species.

plasma

The liquid component of the blood. It is a straw-colored fluid, largely composed of water (around 90 percent), in which a number of substances are dissolved. These include a variety of proteins (around 7 percent) such as fibrinogen (important in blood clotting), inorganic mineral salts such as sodium and calcium, waste products such as urea, traces of hormones, and antibodies to defend against infection.

plasma membrane
See **cell membrane.**

plasmid
Small, mobile piece of DNA found in bacteria and used in genetic engineering. Plasmids are separate from the bacterial chromosome but still multiply during cell growth. Their size ranges from 3 percent to 20 percent of the size of the chromosome. There is usually only one copy of a single plasmid per cell, but occasionally several are found. Some plasmids carry "fertility genes" that enable them to move from one bacterium to another and transfer genetic information between strains. Plasmid genes determine a wide variety of bacterial properties including resistance to antibiotics and the ability to produce toxins.

plastid
General name for a cell organelle of plants that is enclosed by a double membrane and contains a series of internal membranes and vesicles. Plastids contain DNA and are produced by division of existing plastids. They can be classified into two main groups: the chromoplasts, which contain pigments such as carotenes and chlorophyll, and the leucoplasts, which are colorless; however, the distinction between the two is not always clear-cut.

platelet
Tiny disc-shaped structure found in the blood, which helps it to clot. Platelets are not true cells, but membrane-bound cell fragments without nuclei that bud off from large cells in the bone marrow.

plumule
Part of a seed embryo that develops into the shoot, bearing the first true leaves of the plant. In most seeds, for example the sunflower, the plumule is a small conical structure without any leaf structure. Growth of the plumule does not occur until the cotyledons have grown above ground. This is epigeal germination. However, in seeds such as the broad bean, a leaf structure is visible on the plumule in the seed. These seeds develop by the plumule growing up through the soil with the cotyledons remaining below the surface. This is known as hypogeal germination.

pneumatophore
Erect root that rises up above the soil or water and promotes gas exchange. Pneumatophores, or breathing roots, are formed by certain swamp-dwelling trees, such as mangroves, since there is little oxygen available to the roots in waterlogged conditions. They have numerous pores or lenticels over their surface, allowing gas exchange.

pneumothorax

The presence of air in the pleural cavity, between a lung and the chest wall. It may be due to a penetrating injury of the lung or to lung disease, or it may occur without apparent cause (spontaneous pneumothorax) in an otherwise healthy person. Prevented from expanding normally, the lung is liable to collapse.

pod

In botany, a type of fruit that is characteristic of legumes (plants belonging to the Leguminosae family), such as peas and beans. It develops from a single carpel and splits down both sides when ripe to release the seeds.

poikilothermy

The condition in which an animal's body temperature is largely dependent on the temperature of the air or water in which it lives. It is characteristic of all animals except birds and mammals, which maintain their body temperatures by homeothermy (they are "warm-blooded").

pollen

The grains of seed plants that contain the male gametes. In angiosperms (flowering plants) pollen is produced within anthers; in most gymnosperms (cone-bearing plants) it is produced in male cones. A pollen grain is typically yellow and, when mature, has a hard outer wall. Pollen of insect-pollinated plants is often sticky and spiny and larger than the smooth, light grains produced by wind-pollinated species.

pollen tube

Outgrowth from a pollen grain that grows toward the ovule, following germination of the grain on the stigma. In angiosperms (flowering plants) the pollen tube reaches the ovule by growing down through the style, carrying the male gametes inside. The gametes are discharged into the ovule and one fertilizes the egg cell.

pollination

The process by which pollen is transferred from one plant to another. The male gametes are contained in pollen grains, which must be transferred from the anther to the stigma in angiosperms (flowering plants), and from the male cone to the female cone in gymnosperms (cone-bearing plants). Fertilization (not the same as pollination) occurs after the growth of the pollen tube to the ovary. Self-pollination occurs when pollen is transferred to a stigma of the same flower, or to another flower on the same plant; cross-pollination occurs when pollen is transferred to another plant. This involves external pollen-carrying agents, such as wind, water (hydrophily), insects, birds (ornithophily), bats, and other small mammals.

pollution

The harmful effect on the environment of by-products of human activity, principally industrial and agricultural processes—for example, noise, smoke, car emissions, chemical and radioactive effluents in air, seas, and rivers, pesticides, radiation, sewage, and household waste. Pollution contributes to the greenhouse effect.

polymorphism
In genetics, the coexistence of several distinctly different types in a population (groups of animals of one species). Examples include the different blood groups in humans, different color forms in some butterflies, and snail shell size, length, shape, color, and stripiness.

polyp
In zoology, the sedentary stage in the life cycle of a coelenterate (such as a coral or jellyfish), the other being the free-swimming medusa.

polypeptide
Long-chain peptide.

polyploid
In genetics, possessing three or more sets of chromosomes in cases where the normal complement is two sets (diploid). Polyploidy arises spontaneously and is common in plants (mainly among flowering plants), but rare in animals. Many crop plants are natural polyploids, including wheat, which has four sets of chromosomes per cell (durum wheat) or six sets (common wheat). Plant breeders can induce the formation of polyploids by treatment with a chemical, colchicine.

polysaccharide
Long-chain carbohydrate made up of hundreds or thousands of linked simple sugars (monosaccharides) such as glucose and closely related molecules.

polyunsaturate
Type of fat or oil containing a high proportion of triglyceride molecules whose fatty acid chains contain several double bonds. By contrast, the fatty-acid chains of the triglycerides in saturated fats (such as lard) contain only single bonds. Medical evidence suggests that polyunsaturated fats, used widely in margarines and cooking fats, are less likely to contribute to cardiovascular disease than saturated fats, but there is also some evidence that they may have adverse effects on health.

pome
Type of pseudocarp, or false fruit, typical of certain plants belonging to the Rosaceae family. The outer skin and fleshy tissues are developed from the receptacle (the enlarged end of the flower stalk) after fertilization, and the five carpels (the true fruit) form the pome's core, which surrounds the seeds. Examples of pomes are apples, pears, and quinces.

population
In biology and ecology, a group of animals of one species, living in a certain area and able to interbreed; the members of a given species in a community of living things.

population cycle
Regular fluctuations in the size of a population, as seen in lemmings, for example. Such cycles are often caused by density-dependent mortality: high mortality due to overcrowding causes a sudden decline in the population, which then gradually builds up again. Population cycles may also result from an interaction between a predator and its prey.

pore
A pore is a small opening in the skin that releases sweat and sebum. Sebum acts as a natural lubricant and protects the skin from the effects of moisture or excessive dryness.

pregnancy
In humans, the period during which an embryo grows within the womb. It begins at conception and ends at birth, and the normal length is forty weeks. Menstruation usually stops on conception. About one in five pregnancies fails, but most of these failures occur very early on, so the woman may notice only that her period is late. After the second month, the breasts become tense and tender, and the areas round the nipples become darker. Enlargement of the uterus can be felt at about the end of the third month, and thereafter the abdomen enlarges progressively. Fetal movement can be felt at about eighteen weeks; a heartbeat may be heard during the sixth month. Pregnancy in animals is called gestation.

premolar
In mammals, one of the large teeth toward the back of the mouth. In herbivores they are adapted for grinding. In carnivores they may be carnassials. Premolars are present in milk dentition as well as permanent dentition.

preservative
Substance (additive) added to a food in order to inhibit the growth of bacteria, yeasts, molds, and other microorganisms, and therefore extend its shelf life. The term sometimes refers to antioxidants (substances added to oils and fats to prevent their becoming rancid) as well. All preservatives are potentially damaging to health if eaten in sufficient quantity. Both the amount used, and the foods in which they can be used, are restricted by law.

primary sexual characteristic
The endocrine gland producing maleness and femaleness. In males, the primary sexual characteristic is the testis; in females it is the ovary. Both are endocrine

glands that produce hormones responsible for secondary sexual characteristics, such as facial hair and a deep voice in males and breasts in females.

primate

In zoology, any member of the order of mammals that includes monkeys, apes, and humans (together called anthropoids), as well as lemurs, bushbabies, lorises, and tarsiers (together called prosimians). Generally, they have forward-directed eyes, gripping hands and feet, opposable thumbs, and big toes. They tend to have nails rather than claws, with gripping pads on the ends of the digits, all adaptations to the arboreal, climbing mode of life.

prion

Acronym for proteinaceous infectious particle, exceptionally small microorganism, a hundred times smaller than a virus. Composed of protein, and without any detectable amount of nucleic acid (genetic material), it is thought to cause diseases such as scrapie in sheep, and certain degenerative diseases of the nervous system in humans. How it can operate without nucleic acid is not yet known.

productivity, biological

In an ecosystem, the amount of material in the food chain produced by the primary producers (plants) that is available for consumption by animals. Plants turn carbon dioxide and water into sugars and other complex carbon compounds by means of photosynthesis. Their net productivity is defined as the quantity of carbon compounds formed, less the quantity used up by the respiration of the plant itself.

progesterone

Steroid hormone that occurs in vertebrates. In mammals, it regulates the menstrual cycle and pregnancy. Progesterone is secreted by the corpus luteum (the ruptured Graafian follicle of a discharged ovum).

prokaryote

An organism whose cells lack organelles (specialized segregated structures such as nuclei, mitochondria, and chloroplasts). Prokaryote DNA is not arranged in chromosomes but forms a coiled structure called a nucleoid. The prokaryotes comprise only the bacteria and cyanobacteria (blue-green algae); all other organisms are eukaryotes.

proprioceptor

One of the sensory nerve endings that are located in muscles, tendons, and joints. They relay information on the position of the body and the state of muscle contraction.

prop root, or stilt root

Modified root that grows from the lower part of a stem or trunk down to the ground, providing a plant with extra support. Prop roots are common on some woody plants, such as mangroves, and also occur on a few herbaceous plants, such as corn. **Buttress roots** are a type of prop root found at the base of tree trunks, extended and flattened along the upper edge to form massive triangular buttresses; they are common on tropical trees.

prosimian, or primitive primate
In zoology, any animal belonging to the suborder Strepsirhin of primates. Prosimians are characterized by a wet nose with slitlike nostrils, the tip of the nose having a prominent vertical groove. Examples are lemurs, pottos, tarsiers, and the aye-aye.

prostaglandin
Any of a group of complex fatty acids present in the body that act as messenger substances between cells. Effects include stimulating the contraction of smooth muscle (for example, of the womb during birth), regulating the production of stomach acid, and modifying hormonal activity. In excess, prostaglandins may produce inflammatory disorders such as arthritis. Synthetic prostaglandins are used to induce labor in humans and domestic animals.

prostate gland
Gland surrounding and opening into the urethra at the base of the bladder in male mammals.

protandry
In a flower, the state where the male reproductive organs reach maturity before those of the female. This is a common method of avoiding self-fertilization.

protease
General term for a digestive enzyme capable of splitting proteins. Examples include pepsin, found in the stomach, and trypsin, found in the small intestine.

protein
Complex, biologically important substance composed of amino acids joined by peptide bonds. Proteins are essential to all living organisms. As enzymes they regulate all aspects of metabolism. Structural proteins such as keratin and collagen make up the skin, claws, bones, tendons, and ligaments; muscle proteins produce movement; hemoglobin transports oxygen; and membrane proteins regulate the movement of substances into and out of cells. For humans, protein is an essential part of the diet, and is found in greatest quantity in soy beans and other grain legumes, meat, eggs, and cheese.

protein engineering
The creation of synthetic proteins designed to carry out specific tasks. For example, an enzyme may be designed to remove grease from soiled clothes and remain stable at the high temperatures in a washing machine.

amino acids, where R is one of many possible side chains

Peptide – this is one made of just three amino acid units. Proteins consist of very large numbers of amino acid units in long chains, folded up in specific ways.

protein

protein synthesis

Manufacture, within the cytoplasm of the cell, of the proteins an organism needs. The building blocks of proteins are amino acids, of which there are twenty types. The pattern in which the amino acids are linked decides what kind of protein is produced. In turn it is the genetic code, contained within DNA, that determines the precise order in which the amino acids are linked up during protein manufacture. Interestingly, DNA is found only in the nucleus, yet protein synthesis occurs only in the cytoplasm. The information necessary for making the proteins is carried from the nucleus to the cytoplasm by another nucleic acid, RNA.

prothallus

In botany, a short-lived gametophyte of many ferns and other pteridophytes (such as horsetails or club mosses). It bears either the male or female sex organs, or both. Typically it is a small, green, flattened structure that is anchored in the soil by several rhizoids (slender, hairlike structures, acting as roots) and needs damp conditions to survive. The reproductive organs are borne on the lower surface close to the soil.

protist

A single-celled organism which has a eukaryotic cell, but which is not a member of the plant, fungal, or animal kingdoms. The main protists are protozoa.

protogyny

In a flower, the state where the female reproductive organs reach maturity before those of the male. Like protandry, in which the male organs reach maturity first, this is a method of avoiding self-fertilization, but it is much less common.

protoplasm

Contents of a living cell. Strictly speaking it includes all the discrete structures (organelles) in a cell, but it is often used simply to mean the jellylike material in which these float. The contents of a cell outside the nucleus are called cytoplasm.

protozoa

Group of single-celled organisms without rigid cell walls. Some, such as ameba, ingest other cells, but most are saprotrophs or parasites. The group is polyphyletic (containing organisms which have different evolutionary origins).

provitamin

Any precursor substance of a vitamin. Provitamins are ingested substances that become converted to active vitamins within the organism. One example is ergosterol (provitamin D_2), which through the action of sunlight is converted to calciferol (vitamin D_2); another example is beta-carotene, which is hydrolyzed in the liver to vitamin A.

pseudocarp

In botany, a fruitlike structure that incorporates tissue that is not derived from the ovary wall. The additional tissues may be derived from floral parts such as the receptacle and calyx. For example, the colored, fleshy part of a strawberry develops from the receptacle and the true fruits are small achenes—the "pips" embedded in its outer surface. Rose hips are a type of pseudocarp that consists of a hollow, fleshy receptacle containing a number of achenes within. Different types of pseudocarp include pineapples, figs, apples, and pears.

pseudocopulation

Attempted copulation by a male insect with a flower. It results in pollination of the flower and is common in the orchid family, where the flowers of many species resemble a particular species of female bee. When a male bee attempts to mate with a flower, the pollinia (groups of pollen grains) stick to its body. They are transferred to the stigma of another flower when the insect attempts copulation again.

pteridophyte

Simple type of vascular plant. The pteridophytes comprise four classes: the Psilosida, including the most primitive vascular plants, found mainly in the tropics; the Lycopsida, including the club mosses; the Sphenopsida, including the horsetails; and the Pteropsida, including the ferns. They do not produce seeds.

puberty

Stage in human development when the individual becomes sexually mature. It may occur from the age of ten upward. The sexual organs take on their adult form and pubic hair grows. In girls, menstruation begins, and the breasts develop; in boys, the voice breaks and becomes deeper, and facial hair develops.

pubes

Lowest part of the front of the human trunk, the region where the external generative organs are situated. The underlying bony structure, the pubic arch, is formed by the union in the midline of the two pubic bones, which are the front portions of the hip bones. In women this is more prominent than in men, to allow more room for the passage of the child's head at birth, and it carries a pad of fat and connective tissue, the *mons veneris* (mount of Venus), for its protection.

pulse

Impulse transmitted by the heartbeat throughout the arterial systems of vertebrates. When the heart muscle contracts, it forces blood into the aorta (the chief artery). Because the arteries are elastic, the sudden rise of pressure causes a throb or sudden swelling through them. The actual flow of the blood is about 60 cm/2 ft a second in humans. The average adult pulse rate is generally about 70 per minute. The pulse can be felt where an artery is near the surface, for example in the wrist or the neck.

punctuated equilibrium model

Evolutionary theory developed by Niles Eldredge and U.S. paleontologist Stephen Jay Gould in 1972 to explain discontinuities in the fossil record. It claims that periods of rapid change alternate with periods of relative stability (stasis), and that the appearance of new lineages is a separate process from the gradual evolution of adaptive changes within a species.

pupa

Nonfeeding, largely immobile stage of some insect life cycles, in which larval tissues are broken down, and adult tissues and structures are formed.

putrefaction

Decomposition of organic matter by microorganisms.

pyridoxine, or vitamin B$_6$

Water-soluble vitamin ($C_8H_{11}NO_3$) of the B complex. There is no clearly identifiable disease associated with deficiency but its absence from the diet can give rise to malfunction of the central nervous system and general skin disorders. Good sources are liver, meat, milk, and cereal grains. Related compounds may also show vitamin B$_6$ activity.

raceme

In botany, a type of inflorescence.

radial artery

Artery that passes down the forearm and supplies blood to the hand and the fingers. The brachial artery, a large artery supplying blood to the arm, divides at the elbow to form the radial and ulnar arteries. The pulsation of blood through the radial artery can be felt at the wrist. This is generally known as the pulse.

radial nerve

The nerve in the upper arm. Nervous impulses to regulate the function of the muscles which extend the arm, the wrist, and some fingers pass along these nerves. They also relay sensation to parts of the arm and hand. The radial nerve arises from the brachial plexus (network of nerves supplying the arm) in the armpit and descends the upper arm before dividing into the superficial radial and interosseous nerves.

radiation biology

Study of how living things are affected by radioactive (ionizing) emissions and by electromagnetic (nonionizing) radiation (electromagnetic waves). Both are potentially harmful and can cause mutations as well as leukemia and other cancers; even low levels of radioactivity are very dangerous. Both however, are used therapeutically, for example to treat cancer, when the radiation dose is very carefully controlled (radiotherapy or X-ray therapy).

radicle

Part of a plant embryo that develops into the primary root. Usually it emerges from the seed before the embryonic shoot, or plumule, its tip protected by a root cap, or calyptra, as it pushes through the soil. The radicle may form the basis of the entire root system, or it may be replaced by adventitious roots (positioned on the stem).

radius

One of the two bones in the lower forelimb of tetrapod (four-limbed) vertebrates.

receptacle

The enlarged end of a flower stalk to which the floral parts are attached. Normally the receptacle is rounded, but in some plants it is flattened or cup-shaped. The term is also used for the region on that part of some seaweeds which becomes swollen at certain times of the year and bears the reproductive organs.

receptor

Receptors are discrete areas of cell membranes or areas within cells with which neurotransmitters, hormones, and drugs interact. Such interactions control the activities of the body. For example, adrenaline transmits nervous impulses to receptors in the sympathetic nervous system which initiates the characteristic response to excitement and fear in an individual.

recessive gene

In genetics, an allele (alternative form of a gene) that will show in the phenotype (observed characteristics of an organism) only if its partner allele on the paired chromosome is similarly recessive. Such an allele will not show if its partner is dominant, that is if the organism is heterozygous for a particular characteristic. Alleles for blue eyes in humans, and for shortness in pea plants are recessive. Most mutant alleles are recessive and therefore are only rarely expressed.

recombinant DNA

In genetic engineering, DNA formed by splicing together genes from different sources into new combinations.

recombination

In genetics, any process that recombines, or "shuffles," the genetic material, thus increasing genetic variation in the offspring. The two main processes of recombination both occur during meiosis (reduction division of cells). One is **crossing over,** in which chromosome pairs exchange segments; the other is the random reassortment of chromosomes that occurs when each gamete (sperm or egg) receives only one of each chromosome pair.

rectum

Lowest part of the large intestine of animals, which stores feces prior to elimination (defecation).

red blood cell, or erythrocyte

The most common type of blood cell, responsible for transporting oxygen around the body. It contains hemoglobin, which combines with oxygen from the lungs to form oxyhemoglobin. When transported to the tissues, these cells are able to release the oxygen because the oxyhemoglobin splits into its original constituents.

reflex

In animals, a very rapid involuntary response to a particular stimulus. It is controlled by the nervous system. A reflex involves only a few nerve cells, unlike the slower but more complex responses produced by the many processing nerve cells of the brain.

regeneration

Regrowth of a new organ or tissue after the loss or removal of the original. It is common in plants, where a new individual can often be produced from a "cutting" of the original. In animals, regeneration of major structures is limited to lower organisms; certain lizards can regrow their tails if these are lost, and new flatworms can grow from a tiny fragment of an old one. In mammals, regeneration is limited to the repair of tissue in wound healing and the regrowth of peripheral nerves following damage.

REM sleep

Acronym for "rapid-eye-movement" sleep, phase of sleep that recurs several times nightly in humans and is associated with dreaming. The eyes flicker quickly beneath closed lids.

rennin, or chymase

Enzyme found in the gastric juice of young mammals, used in the digestion of milk.

replication

Production of copies of the genetic material DNA; it occurs during cell division (mitosis and meiosis). Most mutations are caused by mistakes during replication.

reproduction

The process by which a living organism produces other organisms more or less similar to itself. The ways in which species reproduce differ, but the two main methods are by asexual reproduction and sexual reproduction. Asexual reproduction involves only one parent without the formation of gametes: the parent's cells divide by mitosis to produce new cells with the same number and kind of chromosomes as its own. Thus offspring produced asexually are clones of the parent and there is no variation. Sexual reproduction involves two parents, one male and one female. The parents' sex cells divide by meiosis producing gametes, which contain only half the number of chromosomes of the parent cell. In this way, when two sets of chromosomes combine during fertilization, a new combination of genes is produced. Hence the new organism will differ from both parents, and variation is introduced. The ability to reproduce is considered one of the fundamental attributes of living things.

reptile

Any member of a class (Reptilia) of vertebrates. Unlike amphibians, reptiles have hard-shelled, yolk-filled eggs that are laid on land and from which fully formed young are born. Some snakes and lizards retain their eggs and give birth to live young. Reptiles are cold-blooded, and their skin is usually covered with scales. The metabolism is slow, and in some cases (certain large snakes) intervals between meals may be months. Reptiles date back over three hundred million years.

respiration

Metabolic process in organisms in which food molecules are broken down to release energy. The cells of all living organisms need a continuous supply of energy, and in most plants and animals this is obtained by aerobic respiration. In this process, oxygen is used to break down the glucose molecules in food. This releases energy in the form of energy-carrying molecules (ATP), and produces carbon dioxide and water as by-products. Respiration sometimes occurs without oxygen, and this is called anaerobic respiration. In this case, the end products are energy and either lactose acid or ethanol (alcohol) and carbon dioxide; this process is termed fermentation.

respiratory surface

Area used by an organism for the exchange of gases, for example the lungs, gills, or, in plants, the leaf interior. The gases oxygen and carbon dioxide are both usually involved in respiration and photosynthesis. Although organisms have evolved different types of respiratory surface according to need, there are certain features in common. These include thinness and moistness, so that the gas can dissolve in a membrane and then diffuse into the body of the organism. In many animals the gas is then transported away from the surface and toward interior cells by the blood system.

restriction enzyme
Bacterial enzyme that breaks a chain of DNA into two pieces at a specific point; used in genetic engineering. The point along the DNA chain at which the enzyme can work is restricted to places where a specific sequence of base pairs occurs. Different restriction enzymes will break a DNA chain at different points. The overlap between the fragments is used in determining the sequence of base pairs in the DNA chain.

retina
Light-sensitive area at the back of the eye connected to the brain by the optic nerve. It has several layers and in humans contains over a million rods and cones, sensory cells capable of converting light into nervous messages that pass down the optic nerve to the brain.

retinol, or vitamin A
Fat-soluble chemical derived from ß-carotene and found in milk, butter, cheese, egg yolk, and liver. Lack of retinol in the diet leads to the eye disease xerophthalmia.

rhesus factor
Group of antigens on the surface of red blood cells of humans which characterize the rhesus blood group system. Most individuals possess the main rhesus factor (Rh+), but those without this factor (Rh–) produce antibodies if they come into contact with it. The name comes from rhesus monkeys, in whose blood rhesus factors were first found.

rhizoid
Hairlike outgrowth found on the gametophyte generation of ferns, mosses, and liverworts. Rhizoids anchor the plant to the substrate and can absorb water and nutrients. They may be composed of many cells, as in mosses, where they are usually brownish, or may be unicellular, as in liverworts, where they are usually colorless. Rhizoids fulfill the same functions as the roots of higher plants but are simpler in construction.

rhizome, or rootstock
Horizontal underground plant stem. It is a perennating organ in some species, where it is generally thick and fleshy, while in other species it is mainly a means of vegetative reproduction, and is therefore long and slender, with buds all along it that send up new plants. The potato is a rhizome that has two distinct parts, the tuber being the swollen end of a long, cordlike rhizome.

rhythm method

Method of natural contraception that relies on refraining from intercourse during ovulation. The time of ovulation can be worked out by the calendar (counting days from the last period), by temperature changes, or by inspection of the cervical mucus. All these methods are unreliable because it is possible for ovulation to occur at any stage of the menstrual cycle.

rib

Long, usually curved bone that extends laterally from the spine in vertebrates. Most fishes and many reptiles have ribs along most of the spine, but in mammals they are found only in the chest area. In humans, there are twelve pairs of ribs. The ribs protect the lungs and heart, and allow the chest to expand and contract easily.

riboflavin, or vitamin B₂

Vitamin of the B complex important in cell respiration. It is obtained from eggs, liver, and milk. A deficiency in the diet causes stunted growth.

ribonucleic acid

Full name of RNA.

ribosome

The protein-making machinery of the cell. Ribosomes are located on the endoplasmic reticulum (ER) of eukaryotic cells, and are made of proteins and a special type of RNA, ribosomal RNA. They receive messenger RNA (copied from the DNA) and amino acids, and "translate" the messenger RNA by using its chemically coded instructions to link amino acids in a specific order, to make a strand of a particular protein.

rickets

Defective growth of bone in children due to an insufficiency of calcium deposits. The bones, which do not harden adequately, are bent out of shape. It is usually caused by a lack of vitamin D and insufficient exposure to sunlight. Renal rickets, also a condition of malformed bone, is associated with kidney disease.

ritualization

In ethology, a stereotype that occurs in certain behavior patterns when these are incorporated into displays. For example, the exaggerated and stylized head toss of the goldeneye drake during courtship is a ritualization of the bathing movement used to wet the feathers; its duration and form have become fixed. Ritualization may make displays clearly recognizable, so ensuring that individuals mate only with members of their own species.

RNA

Abbreviation for ribonucleic acid, nucleic acid involved in the process of translating the genetic material DNA into proteins. It is usually single-stranded, unlike the double-stranded DNA, and consists of a large number of nucleotides strung together, each of which comprises the sugar ribose, a phosphate group, and one of four bases (uracil, cytosine, adenine, or guanine). RNA is copied from DNA by the formation of base pairs, with uracil taking the place of thymine.

rodent
Any mammal of the worldwide order Rodentia, making up nearly half of all mammal species. Besides ordinary "cheek teeth," they have a single front pair of incisor teeth in both upper and lower jaw, which continue to grow as they are worn down.

root
The part of a plant that is usually underground, and whose primary functions are anchorage and the absorption of water and dissolved mineral salts. Roots usually grow downward and toward water (that is, they are positively geotropic and hydrotropic). Plants such as epiphytic orchids, which grow above ground, produce aerial roots that absorb moisture from the atmosphere. Others, such as ivy, have climbing roots arising from the stems, which serve to attach the plant to trees and walls.

root hair
Tiny hairlike outgrowth on the surface cells of plant roots that greatly increases the area available for the absorption of water and other materials. It is a delicate structure, which survives for a few days only and does not develop into a root.

root nodule
Clearly visible swelling that develops in the roots of members of the bean family, the Leguminosae. The cells inside this tumorous growth have been invaded by the bacteria Rhizobium, a soil microbe capable of converting gaseous nitrogen into nitrate. The nodule is therefore an association between a plant and a bacterium, with both partners benefiting. The plant obtains nitrogen compounds while the bacterium obtains nutrition and shelter.

rootstock
Another name for rhizome, an underground plant organ.

rotifer
Any of the tiny invertebrates, also called "wheel animalcules," of the phylum Rotifera. Mainly freshwater, some marine, rotifers have a ring of cilia that carries food to the mouth and also provides propulsion. They are the smallest of multicellular animals—few reach 0.05 cm/0.02 in.

roughage
See fiber, dietary.

ruminant
Any even-toed hoofed mammal with a rumen, the "first stomach" of its complex digestive system. Plant food is stored and fermented before being brought back to the mouth for chewing (chewing the cud) and then is swallowed to the next stomach. Ruminants include cattle, antelopes, goats, deer, and giraffes, all with a four-chambered stomach. Camels are also ruminants, but they have a three-chambered stomach.

runner, or stolon
In botany, aerial stem that produces new plants.

saccharide
Another name for a sugar molecule.

saliva
In vertebrates, an alkaline secretion from the salivary glands that aids the swallowing and digestion of food in the mouth. In mammals, it contains the enzyme amylase, which converts starch to sugar. The salivary glands of mosquitoes and other blood-sucking insects produce anticoagulants.

salivary gland, or parotid gland
In mammals, one of two glands situated near the mouth responsible for the manufacture of saliva and its secretion into the mouth. The salivary glands are stimulated to produce saliva during a meal. Saliva contains an enzyme, ptyalin, and mucous which are essential for the mastication and initial digestion of food.

salmonella
Any of a very varied group of bacteria, genus *Salmonella* that colonize the intestines of humans and some animals. Some strains cause typhoid and paratyphoid fevers, while others cause salmonella food poisoning, which is characterized by stomach pains, vomiting, diarrhea, and headache. It can be fatal in elderly people, but others usually recover in a few days without antibiotics. Most cases are caused by contaminated animal products, especially poultry meat.

samara
In botany, a winged fruit, a type of achene.

sap
The fluids that circulate through vascular plants, especially woody ones. Sap carries water and food to plant tissues. Sap contains alkaloids, protein, and starch; it can be milky (as in rubber trees), resinous (as in pines), or syrupy (as in maples).

saprotroph
Formerly saprophyte, organism that feeds on the excrement or the dead bodies or tissues of others. They include most fungi (the rest being parasites); many

bacteria and protozoa; animals such as dung beetles and vultures; and a few unusual plants, including several orchids. Saprotrophs cannot make food for themselves, so they are a type of heterotroph. They are useful scavengers, and in sewage farms and refuse dumps break down organic matter into nutrients easily assimilable by green plants.

sarcoma
Malignant tumor arising from the fat, muscles, bones, cartilage, or blood and lymph vessels and connective tissues. Sarcomas are much less common than carcinomas.

saturated fatty acid
Fatty acid in which there are no double bonds in the hydrocarbon chain.

scapula
Large, flat, triangular bone (the "shoulder blade") that lies over the second to seventh ribs on the back, forming part of the pectoral girdle and assisting in the articulation of the arm with the chest region. Its flattened shape allows a large region for the attachment of muscles.

scent gland
Gland that opens onto the outer surface of animals, producing odorous compounds that are used for communicating between members of the same species (pheromones), or for discouraging predators.

schizocarp
Dry fruit that develops from two or more carpels and splits, when mature, to form separate one-seeded units known as mericarps.

sclerenchyma
Plant tissue whose function is to strengthen and support, composed of thick-walled cells that are heavily lignified (toughened). On maturity the cell inside dies, and only the cell walls remain.

scurvy
Disease caused by deficiency of vitamin C (ascorbic acid), which is contained in fresh vegetables and fruit. The signs are weakness and aching joints and muscles, progressing to bleeding of the gums and other spontaneous hemorrhage, and drying-up of the skin and hair. It is reversed by giving the vitamin.

sea anemone
Invertebrate marine animal of the phylum Cnidaria with a tubelike body attached by the base to a rock or shell. The other end has an open "mouth" surrounded by stinging tentacles, which capture crustaceans and other small organisms. Many sea anemones are beautifully colored, especially those in tropical waters.

sebum
Oily secretion from the sebaceous glands that acts as a skin lubricant. Acne is caused by inflammation of the sebaceous glands and oversecretion of sebum.

secondary growth, or secondary thickening
Increase in diameter of the roots and stems of certain plants (notably shrubs and trees) that results from the production of new cells by the cambium. It provides the plant with additional mechanical support and new conducting cells, the secondary xylem and phloem. Secondary growth is generally confined to gymnosperms and, among the angiosperms, to the dicotyledons. With just a few exceptions, the monocotyledons (grasses, lilies) exhibit only primary growth, resulting from cell division at the apical meristems.

secondary sexual characteristic
An external feature of an organism that is indicative of its gender (male or female), but not the reproductive organs themselves. They include facial hair in men and breasts in women, combs in roosters, brightly colored plumage in many male birds, and manes in male lions. In many cases, they are involved in displays and contests for mates and have evolved by sexual selection. Their development is stimulated by sex hormones.

secretin
Hormone produced by the small intestine of vertebrates that stimulates the production of digestive secretions by the pancreas and liver.

secretion
Any substance (normally a fluid) produced by a cell or specialized gland, for example, sweat, saliva, enzymes, and hormones. The process whereby the substance is discharged from the cell is also known as secretion.

sedative
Any drug that has a calming effect, reducing anxiety and tension. Sedatives will induce sleep in larger doses. Examples are barbiturates, narcotics, and benzodiazepines.

seed
The reproductive structure of higher plants (angiosperms and gymnosperms). It develops from a fertilized ovule and consists of an embryo and a food store, surrounded and protected by an outer seed coat, called the testa. The food store is contained either in a specialized nutritive tissue, the endosperm, or in the cotyledons of the embryo itself. In angiosperms the seed is enclosed within a fruit, whereas in gymnosperms it is usually naked and unprotected, once shed from the female cone. Following germination the seed develops into a new plant.

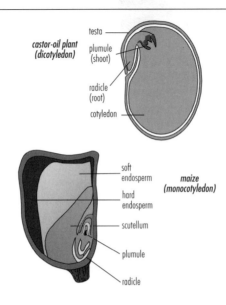

seed

seed plant

Any seed-bearing plant; also known as a spermatophyte. The seed plants are sub-divided into two classes: the angiosperms, or flowering plants, and the gymno-sperms, principally the cycads and conifers. Together, they comprise the major types of vegetation found on land.

semelparity

The occurrence of a single act of reproduction during an organism's lifetime. Most semelparous species produce very large numbers of offspring when they do re-produce, and normally die soon afterwards. Examples include the Pacific salmon and the pine looper moth. Many plants are semelparous, or monocarpic. Repeated reproduction is called iteroparity.

semen

Fluid containing sperm from the testes and secretions from various sex glands (such as the prostate gland) that is ejaculated by male animals during copulation. The secretions serve to nourish and activate the sperm cells, and prevent them clumping together.

sense organ

Any organ that an animal uses to gain information about its surroundings. All sense organs have specialized receptors (such as light receptors in the eye) and some means of translating their response into a nerve impulse that travels to the brain. The main human sense organs are the eye, which detects light and color (different wavelengths of light); the ear, which detects sound (vibrations of the air) and gravity; the nose, which detects some of the chemical molecules in the air; and the tongue, which detects some of the chemicals in food, giving a sense of taste. There are also many small sense organs in the skin, including pain, temperature, and pressure sensors, contributing to our sense of touch.

sensitivity

The ability of an organism, or part of an organism, to detect changes in the environment. All living things are capable of some sensitivity, and any change detected by an organism is called a stimulus. Plant response to stimuli (for example, light, heat, moisture) is by directional growth (tropism). In animals, the body cells that detect the stimuli are called receptors, and these are often contained within a sense organ. For example, the eye is a sense organ, within which the retina contains rod and cone cells which are receptors. The part of the body that responds to a stimulus, such as a muscle, is called an effector, and the communication of stimuli from receptors to effectors is termed "coordination"; messages are passed from receptors to effectors either via the nerves or by means of chemicals called hormones. Rapid communication and response to stimuli, such as light, sound, and scent, can be essential to an animal's well-being and survival, and evolution has led to the development of highly complex mechanisms for this purpose.

sepal

Part of a flower, usually green, that surrounds and protects the flower in bud. The sepals are derived from modified leaves, and are collectively known as the calyx.

sequencing

In biochemistry, determining the sequence of chemical subunits within a large molecule. Techniques for sequencing amino acids in proteins were established in the 1950s, insulin being the first for which the sequence was completed. The Human Genome Project is attempting to determine the sequence of the three billion base pairs within human DNA.

sere

Plant succession developing in a particular habitat. A lithosere is a succession starting on the surface of bare rock. A hydrosere is a succession in shallow freshwater, beginning with planktonic vegetation and the growth of pondweeds and other aquatic plants, and ending with the development of swamp. A plagiosere is the sequence of communities that follows the clearing of the existing vegetation.

serum

Clear fluid that separates out from clotted blood. It is blood plasma with the anti-coagulant proteins removed, and contains antibodies and other proteins, as well as the fats and sugars of the blood. It can be produced synthetically, and is used to protect against disease.

sessile

In botany, a leaf, flower, or fruit that lacks a stalk and sits directly on the stem, as with the sessile acorns of certain oaks. In zoology, it is an animal that normally stays in the same place, such as a barnacle or mussel. The term is also applied to the eyes of crustaceans when these lack stalks and sit directly on the head.

sex determination

Process by which the sex of an organism is determined. In many species, the sex of an individual is dictated by the two sex chromosomes (X and Y) it receives from its parents. In mammals, some plants, and a few insects, males are XY, and females XX; in birds, reptiles, some amphibians, and butterflies the reverse is the case. In bees and wasps, males are produced from unfertilized eggs, females from fertilized eggs. Environmental factors can affect some fish and reptiles, such as turtles, where sex is influenced by the temperature at which the eggs develop. In 1991 it was shown that maleness is caused by a single gene, fourteen base pairs long, on the Y chromosome.

sex hormone

Steroid hormone produced and secreted by the gonads (testes and ovaries). Sex hormones control development and reproductive functions and influence sexual and other behavior.

sex linkage

In genetics, the tendency for certain characteristics to occur exclusively, or predominantly, in one sex only. Human examples include red-green color blindness and hemophilia, both found predominantly in males. In both cases, these characteristics are recessive and are determined by genes on the X chromosome.

sexual reproduction

Reproductive process in organisms that requires the union, or fertilization, of gametes (such as eggs and sperm). These are usually produced by two different individuals, although self-fertilization occurs in a few hermaphrodites such as tapeworms. Most organisms other than bacteria and cyanobacteria (blue-green algae) show some sort of sexual process. Except in some lower organisms, the gametes are of two distinct types called eggs and sperm. The organisms producing the eggs are called females, and those producing the sperm, males. The fusion of a male and female gamete produces a zygote, from which a new individual develops.

sexual selection

Process similar to natural selection but relating exclusively to success in finding a mate for the purpose of sexual reproduction and producing offspring. Sexual selection occurs when one sex (usually but not always the female) invests more effort in producing young than the other. Members of the other sex compete for access to this limited resource (usually males competing for the chance to mate with females).

shell
The hard outer covering of a wide variety of invertebrates. The covering is usually mineralized, normally with large amounts of calcium. The shell of birds' eggs is also largely made of calcium.

shoot
In botany, the parts of a vascular plant growing above ground, comprising a stem bearing leaves, buds, and flowers. The shoot develops from the plumule of the embryo.

shrub
Perennial woody plant that typically produces several separate stems, at or near ground level, rather than the single trunk of most trees. A shrub is usually smaller than a tree, but there is no clear distinction between large shrubs and small trees.

sickle-cell disease
Hereditary chronic blood disorder common among people of black African descent; also found in the eastern Mediterranean, parts of the Persian Gulf, and in northeast India. It is characterized by distortion and fragility of the red blood cells, which are lost too rapidly from the circulation. This often results in anemia.

simple sugar
See **monosaccharide.**

siphonogamy
Plant reproduction in which a pollen tube grows to enable male gametes to pass to the ovary without leaving the protection of the plant. Siphonogamous reproduction is found in all angiosperms and most gymnosperms, and has enabled these plants to reduce their dependency on wet conditions for reproduction, unlike the zoidogamous plants.

skeleton
The rigid or semirigid framework that supports and gives form to an animal's body, protects its internal organs, and provides anchorage points for its muscles. The skeleton may be composed of bone and cartilage (vertebrates), chitin (arthropods), calcium carbonate (mollusks and other invertebrates), or silica (many protists). The human skeleton is composed of 206 bones, with the vertebral column (spine) forming the central supporting structure.

skin
The covering of the body of a vertebrate. In mammals, the outer layer (epidermis) is dead and its cells are constantly being rubbed away and replaced from below; it helps to protect the body from infection and to prevent dehydration. The lower layer (dermis) contains blood vessels, nerves, hair roots, and sweat and

sebaceous glands, and is supported by a network of fibrous and elastic cells. The medical specialty concerned with skin diseases is called dermatology.

skull

In vertebrates, the collection of flat and irregularly shaped bones (or cartilage) that enclose the brain and the organs of sight, hearing, and smell, and provide support for the jaws. In most mammals, the skull consists of twenty-two bones joined by fibrous immobile joints called sutures. The floor of the skull is pierced by a large hole *(foramen magnum)* for the spinal cord and a number of smaller apertures through which other nerves and blood vessels pass.

sleep

State of natural unconsciousness and activity that occurs at regular intervals in most mammals and birds, though there is considerable variation in the amount of time spent sleeping. Sleep differs from hibernation in that it occurs daily rather than seasonally, and involves less drastic reductions in metabolism. The function of sleep is unclear. People deprived of sleep become irritable, uncoordinated, forgetful, hallucinatory, and even psychotic.

slime mold, or myxomycete

Extraordinary organism that shows some features of fungus and some of protozoa. Slime molds are not closely related to any other group, although they are often classed, for convenience, with the fungi. There are two kinds, cellular slime molds and plasmodial slime molds, differing in their complex life cycles.

smell

Sense that responds to chemical molecules in the air. It works by having receptors for particular chemical groups, into which the airborne chemicals must fit to trigger a message to the brain.

smooth muscle

Involuntary muscle capable of slow contraction over a period of time. It is present in hollow organs, such as the intestines, stomach, bladder, and blood vessels. Its presence in the wall of the alimentary canal allows slow rhythmic movements known as peristalsis, which cause food to be mixed and forced along the gut. Smooth muscle has a microscopic structure distinct from other forms.

social behavior

In zoology, behavior concerned with altering the behavior of other individuals of the same species. Social behavior allows animals to live harmoniously in groups by establishing hierarchies of dominance to discourage disabling fighting. It may be aggressive or submissive (for example, cowering and other signals of appeasement), or designed to establish bonds (such as social grooming or preening).

sociobiology
Study of the biological basis of all social behavior, including the application of population genetics to the evolution of behavior. It builds on the concept of inclusive fitness, contained in the notion of the "selfish gene." Contrary to some popular interpretations, it does not assume that all behavior is genetically determined.

sorus
In ferns, a group of sporangia, the reproductive structures that produce spores. They occur on the lower surface of fern fronds.

spadix
In botany, an inflorescence consisting of a long, fleshy axis bearing many small, stalkless flowers. It is partially enclosed by a large bract or spathe. A spadix is characteristic of plants belonging to the family Araceae, including the arum lily *Zantedeschia aethiopica.*

spathe
In flowers, the single large bract surrounding the type of inflorescence known as a spadix. It is sometimes brightly colored and petal-like, as in the brilliant scarlet spathe of the flamingo plant *Anthurium andreanum* from South America; this serves to attract insects.

speciation
Emergence of a new species during evolutionary history. One cause of speciation is the geographical separation of populations of the parent species, followed by reproductive isolation and selection for different environments so that they no longer produce viable offspring when they interbreed. Other causes are assortative mating and the establishment of a polyploid population.

species
A distinguishable group of organisms that resemble each other or consist of a few distinctive types (as in polymorphism), and that can all interbreed to produce fertile offspring. Species are the lowest level in the system of biological classification.

sperm, or spermatozoon
The male gamete of animals. Each sperm cell has a head capsule containing a nucleus, a middle portion containing mitochondria (which provide energy), and a long tail (flagellum).

spermatophore
Small capsule containing sperm and other nutrients produced in invertebrates, newts, and cephalopods.

spermatophyte

In botany, another name for a seed plant.

sphincter

Ring of muscle, such as is found at various points in the alimentary canal, that contracts and relaxes to open and close the canal and control the movement of food. The pyloric sphincter, at the base of the stomach, controls the release of the gastric contents into the duodenum. After release the sphincter contracts, closing off the stomach. The external anal sphincter closes the anus; the internal anal sphincter constricts the rectum; the sphincter vesicae controls the urethral orifice of the bladder. In the eye the sphincter pupillae contracts the pupil in response to bright light.

spikelet

In botany, one of the units of a grass inflorescence. It comprises a slender axis on which one or more flowers are borne.

spinal cord

Major component of the central nervous system in vertebrates, encased in the spinal column. It consists of bundles of nerves enveloped in three layers of membrane (the meninges).

spine

Backbone of vertebrates. In most mammals, it contains twenty-six small bones called vertebrae, which enclose and protect the spinal cord (which links the peripheral nervous system to the brain). The spine articulates with the skull, ribs, and hip bones, and provides attachment for the back muscles.

spiracle

In insects, the opening of a trachea, through which oxygen enters the body and carbon dioxide is expelled. In cartilaginous fishes (sharks and rays), the same name is given to a circular opening that marks the remains of the first gill slit.

spirochaete

Spiral-shaped bacterium. Some spirochaetes are free-living in water, others inhabit the intestines and genital areas of animals. The sexually transmitted disease syphilis is caused by a spirochaete.

spleen

Organ in vertebrates, part of the reticuloendothelial system, which helps to process lymphocytes. It also regulates the number of red blood cells in circulation by destroying old cells, and stores iron. It is situated on the left side of the body, behind the stomach.

spontaneous generation, or abiogenesis

Erroneous belief that living organisms can arise spontaneously from nonliving matter. This survived until the mid-nineteenth century, when the French chemist Louis Pasteur demonstrated that a nutrient broth would not generate microorganisms if it was adequately sterilized. The theory of biogenesis holds that spontaneous generation cannot now occur; it is thought, however, to have played an essential role in the origin of life on this planet four billion years ago.

sporangium

Structure in which spores are produced.

spore

Small reproductive or resting body, usually consisting of just one cell. Unlike a gamete, it does not need to fuse with another cell in order to develop into a new organism. Spores are produced by the lower plants, most fungi, some bacteria, and certain protozoa. They are generally light and easily dispersed by wind movements. Plant spores are haploid and are produced by the sporophyte, following meiosis.

sporophyte

Diploid spore-producing generation in the life cycle of a plant that undergoes alternation of generations.

stamen

Male reproductive organ of a flower. The stamens are collectively referred to as the androecium. A typical stamen consists of a stalk, or filament, with an anther, the pollen-bearing organ, at its apex, but in some primitive plants, such as *Magnolia*, the stamen may not be markedly differentiated.

staphylococcus

Spherical bacterium that occurs in clusters. It is found on the skin and mucous membranes of humans and other animals. It can cause abscesses and systemic infections that may prove fatal.

starch

Widely distributed, high-molecular-mass carbohydrate, produced by plants as a food store; main dietary sources are cereals, legumes, and tubers, including potatoes. It consists of varying proportions of two glucose polymers (polysaccharides): straight-chain (amylose) and branched (amylopectin) molecules.

stem

Main supporting axis of a plant that bears the leaves, buds, and reproductive structures; it may be simple or branched. The plant stem usually grows above ground, although some grow underground, including rhizomes, corms, rootstocks, and tubers. Stems contain a continuous vascular system that conducts water and food to and from all parts of the plant.

stereocilium
Sensory hair cell found on the cochlea in the inner ear. When the cochlear fluid vibrates in response to sound the hairs move together pushing on protein molecules on their surface. On each stereocilium the protein molecule triggers the opening of an ion gate that enables the flow of potassium ions through the hair. The ion flow causes an electrical signal to be transmitted to the brain as a nerve impulse, where it is interpreted as sound.

sterilization
The killing or removal of living organisms such as bacteria and fungi. A sterile environment is necessary in medicine, food processing, and some scientific experiments. Methods include heat treatment (such as boiling), the use of chemicals (such as disinfectants), irradiation with gamma rays, and filtration.

sternum
The large flat bone (the "breastbone"), 15–20 cm/5.9–7.8 in long in the adult, at the front of the chest, joined to the ribs. It gives protection to the heart and lungs. During open-heart surgery the sternum must be split to give access to the thorax.

steroid
Any of a group of cyclic, unsaturated alcohols (lipids without fatty acid components), which, like sterols, have a complex molecular structure consisting of four carbon rings. Steroids include the sex hormones, such as testosterone, the corticosteroid hormones produced by the adrenal glands, bile acids, and cholesterol. The term is commonly used to refer to anabolic steroid. In medicine, synthetic steroids are used to treat a wide range of conditions.

sterol
Any of a group of solid, cyclic, unsaturated alcohols, with a complex structure that includes four carbon rings; cholesterol is an example. Steroids are derived from sterols.

stigma
In a flower, the surface at the tip of a carpel that receives the pollen. It often has short outgrowths, flaps, or hairs to trap pollen and may produce a sticky secretion to which the grains adhere.

stipule
Outgrowth arising from the base of a leaf or leaf stalk in certain plants. Stipules usually occur in pairs or fused into a single semicircular structure.

stolon

In botany, a type of runner.

stoma

Plural stomata, in botany, a pore in the epidermis of a plant. Each stoma is surrounded by a pair of guard cells that are crescent-shaped when the stoma is open but can collapse to an oval shape, thus closing off the opening between them. Stomata allow the exchange of carbon dioxide and oxygen (needed for photosynthesis and respiration) between the internal tissues of the plant and the outside atmosphere. They are also the main route by which water is lost from the plant, and they can be closed to conserve water, the movements being controlled by changes in turgidity of the guard cells.

stomach

The first cavity in the digestive system of animals. In mammals it is a bag of muscle situated just below the diaphragm. Food enters it from the esophagus, is digested by the acid and enzymes secreted by the stomach lining, and then passes into the duodenum. Some plant-eating mammals have multichambered stomachs that harbor bacteria in one of the chambers to assist in the digestion of cellulose. The gizzard is part of the stomach in birds.

stridulatory organs

In insects, organs that produce sound when rubbed together. Crickets rub their wings together, but grasshoppers rub a hind leg against a wing. Stridulation is thought to be used for attracting mates, but may also serve to mark territory.

strobilus

In botany, a reproductive structure found in most gymnosperms and some pteridophytes, notably the club mosses. In conifers the strobilus is commonly known as a cone.

style

In flowers, the part of the carpel bearing the stigma at its tip. In some flowers it is very short or completely lacking, while in others it may be long and slender, positioning the stigma in the most effective place to receive the pollen.

substrate

In biochemistry, a compound or mixture of compounds acted on by an enzyme. The term also refers to a substance such as agar that provides the nutrients for the metabolism of microorganisms. Since the enzyme systems of microorganisms regulate their metabolism, the essential meaning is the same.

succession

In ecology, a series of changes that occur in the structure and composition of the vegetation in a given area from the time it is first colonized by plants (primary succession), or after it has been disturbed by fire, flood, or clearing (secondary succession).

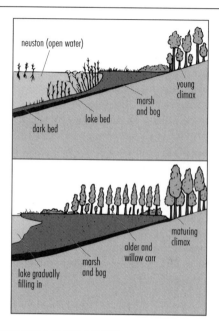

succulent plant
Thick, fleshy plant that stores water in its tissues; for example, cacti and stonecrops *Sedum*. Succulents live either in areas where water is very scarce, such as deserts, or in places where it is not easily obtainable because of the high concentrations of salts in the soil, as in salt marshes. Many desert plants are xerophytes.

suckering
In plants, reproduction by new shoots (suckers) arising from an existing root system rather than from seed. Plants that produce suckers include elm, dandelion, and members of the rose family.

sucrose, or cane sugar or beet sugar
A sugar($C_{12}H_{22}O_{11}$) found in the pith of sugar cane and in sugar beets. It is popularly known as sugar.

suprarenal glands
Alternative name for the adrenal glands.

surface-area-to-volume ratio
The ratio of an animal's surface area (the area covered by its skin) to its total volume. This is high for small animals, but low for large animals such as elephants.

suspensory ligament

In the eye, a ring of fiber supporting the lens. The ligaments attach to the ciliary muscles, the circle of muscle mainly responsible for changing the shape of the lens during accommodation. If the ligaments are put under tension, the lens becomes flatter, and therefore able to focus on objects in the far distance.

sustained-yield cropping

In ecology, the removal of surplus individuals from a population of organisms so that the population maintains a constant size. This usually requires selective removal of animals of all ages and both sexes to ensure a balanced population structure. Taking too many individuals can result in a population decline, as in overfishing.

sweat gland

Gland within the skin of mammals that produces surface perspiration. In primates, sweat glands are distributed over the whole body, but in most other mammals they are more localized; for example, in cats and dogs they are restricted to the feet and around the face.

swim bladder

Thin-walled, air-filled sac found between the gut and the spine in bony fishes. Air enters the bladder from the gut or from surrounding capillaries and changes of air pressure within the bladder maintain buoyancy whatever the water depth.

symbiosis or mutualism

Any close relationship between two organisms of different species, and one where both partners benefit from the association. A well-known example is the pollination relationship between insects and flowers, where the insects feed on nectar and carry pollen from one flower to another. This is sometimes known as mutualism.

synapse

Junction between two nerve cells, or between a nerve cell and a muscle (a neuromuscular junction), across which a nerve impulse is transmitted. The two cells are separated by a narrow gap called the synaptic cleft. The gap is bridged by a chemical neurotransmitter, released by the nerve impulse.

synovial fluid

Viscous colorless fluid that bathes movable joints between the bones of vertebrates. It nourishes and lubricates the cartilage at the end of each bone.

syrinx

The voice-producing organ of a bird. It is situated where the trachea divides in two and consists of vibrating membranes, a reverberating capsule, and numerous controlling muscles.

systole

The contraction of the heart. It alternates with diastole, the resting phase of the heartbeat.

taproot

In botany, a single, robust, main root that is derived from the embryonic root, or radicle, and grows vertically downward, often to considerable depth. Taproots are often modified for food storage and are common in biennial plants such as the carrot *Daucus carota,* where they act as perennating organs.

taste

Sense that detects some of the chemical constituents of food. The human tongue can distinguish only four basic tastes (sweet, sour, bitter, and salty) but it is supplemented by the sense of smell. What we refer to as taste is really a composite sense made up of both taste and smell.

taxis or tactic movement

Plural *taxes,* in botany, the movement of a single cell, such as a bacterium, protozoan, single-celled alga, or gamete, in response to an external stimulus. A movement directed toward the stimulus is described as positive taxis, and away from it as negative taxis. The alga *Chlamydomonas,* for example, demonstrates positive phototaxis by swimming toward a light source to increase the rate of photosynthesis. Chemotaxis is a response to a chemical stimulus, as seen in many bacteria that move toward higher concentrations of nutrients.

taxonomy

Another name for the classification of living organisms.

T cell, or T lymphocyte

Immune cell that plays several roles in the body's defenses. T cells are so called because they mature in the thymus.

tears

Salty fluid exuded by lacrimal glands in the eyes. The fluid contains proteins that are antibacterial, and also absorbs oils and mucus. Apart from cleaning and disinfecting the surface of the eye, the fluid supplies nutrients to the cornea, which does not have a blood supply.

temperature regulation

The ability of an organism to control its internal body temperature. Animals that rely on their environment for their body temperature (and therefore have a variable temperature) are known as ectotherms or "cold-blooded" animals (e.g., lizard). Animals with a constant body temperature irrespective of their environment are known as endotherms or "warm-blooded" animals (e.g., birds and mammals). Their temperature is regulated by the medulla in the brain.

tendon, or sinew
In vertebrates, a cord of very strong, fibrous connective tissue that joins muscle to bone. Tendons are largely composed of bundles of fibers made of the protein collagen, and because of their inelasticity are very efficient at transforming muscle power into movement.

tendril
In botany, a slender, threadlike structure that supports a climbing plant by coiling around suitable supports, such as the stems and branches of other plants. It may be a modified stem, leaf, leaflet, flower, leaf stalk, or stipule (a small appendage on either side of the leaf stalk), and may be simple or branched. The tendrils of Virginia creeper *Parthenocissus quinquefolia* are modified flower heads with suckerlike pads at the end that stick to walls, while those of the grapevine *Vitis* grow away from the light and thus enter dark crevices where they expand to anchor the plant firmly.

territorial behavior
Any behavior that serves to exclude other members of the same species from a fixed area or territory. It may involve aggressively driving out intruders, marking the boundary (with dung piles or secretions from special scent glands), conspicuous visual displays, characteristic songs, or loud calls.

territory
In animal behavior, a fixed area from which an animal or group of animals excludes other members of the same species. Animals may hold territories for many different reasons; for example, to provide a constant food supply, to monopolize potential mates, or to ensure access to refuges or nest sites. The size of a territory depends in part on its function: some nesting and mating territories may be only a few square meters, whereas feeding territories may be as large as hundreds of square kilometers.

testa
The outer coat of a seed, formed after fertilization of the ovule. It has a protective function and is usually hard and dry. In some cases the coat is adapted to aid dispersal, for example by being hairy. Humans have found uses for many types of testa, including the fiber of the cotton seed.

test cross
In genetics, a breeding experiment used to discover the genotype of an individual organism. By crossing with a double recessive of the same species, the offspring will indicate whether the test individual is homozygous or heterozygous for the characteristic in question. In peas, a tall plant under investigation would be crossed with a double recessive short plant with known genotype tt. The results of the cross will be all tall plants if the test plant is TT. If the individual is in fact Tt then there will be some short plants (genotype tt) among the offspring.

testis

Plural *testes,* the organ that produces sperm in male (and hermaphrodite) animals. In vertebrates it is one of a pair of oval structures that are usually internal, but in mammals (other than elephants and marine mammals), the paired testes (or testicles) descend from the body cavity during development, to hang outside the abdomen in a scrotal sac. The testes also secrete the male sex hormone androgen.

testosterone

In vertebrates, hormone secreted chiefly by the testes, but also by the ovaries and the cortex of the adrenal glands. It promotes the development of secondary sexual characteristics in males. In animals with a breeding season, the onset of breeding behavior is accompanied by a rise in the level of testosterone in the blood.

tetrapod

From Greek, meaning "four-legged," type of vertebrate. The group includes mammals, birds, reptiles, and amphibians. Birds are included because they evolved from four-legged ancestors, the forelimbs having become modified to form wings. Even snakes are tetrapods, because they are descended from four-legged reptiles.

thallus

Any plant body that is not divided into true leaves, stems, and roots. It is often thin and flattened, as in the body of a seaweed, lichen, or liverwort, and the gametophyte generation (prothallus) of a fern.

thiamine, or vitamin B₁

A water-soluble vitamin of the B complex. It is found in seeds and grain. Its absence from the diet causes the disease beriberi.

thorax

In four-limbed vertebrates, the part of the body containing the heart and lungs, and protected by the rib cage; in arthropods, the middle part of the body, between the head and abdomen.

throat

In human anatomy, the passage that leads from the back of the nose and mouth to the trachea and esophagus. It includes the pharynx and the larynx, the latter being at the top of the trachea. The word "throat" is also used to mean the front part of the neck, both in humans and other vertebrates; for example, in describing the plumage of birds. In engineering, it is any narrowing entry, such as the throat of a carburetor.

thymus

Organ in vertebrates, situated in the upper chest cavity in humans. The thymus processes lymphocyte cells to produce T-lymphocytes (T denotes "thymus-derived"), which are responsible for binding to specific invading organisms and killing them or rendering them harmless.

thyroid
Endocrine gland of vertebrates, situated in the neck in front of the trachea. It secretes several hormones, principally thyroxine, an iodine-containing hormone that stimulates growth, metabolism, and other functions of the body. The thyroid gland may be thought of as the regulator gland of the body's metabolic rate. If it is overactive, as in hyperthyroidism, the sufferer feels hot and sweaty, has an increased heart rate, diarrhea, and weight loss. Conversely, an underactive thyroid leads to myxedema, a condition characterized by sensitivity to the cold, constipation, and weight gain. In infants, an underactive thyroid leads to cretinism, a form of mental retardation.

thyroxine
In medicine, a hormone containing iodine that is produced by the thyroid gland. It is used to treat conditions that are due to deficiencies in thyroid function, such as myxedema.

tibia
The anterior of the pair of bones in the leg between the ankle and the knee. In humans, the tibia is the shinbone. It articulates with the femur above to form the knee joint, the fibula externally at its upper and lower ends, and with the talus below, forming the ankle joint.

tissue
Any kind of cellular fabric that occurs in an organism's body. Several kinds of tissue can usually be distinguished, each consisting of cells of a particular kind bound together by cell walls (in plants) or extracellular matrix (in animals). Thus, nerve and muscle are different kinds of tissue in animals, as are parenchyma and sclerenchyma in plants.

tissue culture
Process by which cells from a plant or animal are removed from the organism and grown under controlled conditions in a sterile medium containing all the necessary nutrients. Tissue culture can provide information on cell growth and differentiation, and is also used in plant propagation and drug production.

toadstool
Common name for many umbrella-shaped fruiting bodies of fungi. The term is normally applied to those that are inedible or poisonous.

tocopherol, or vitamin E
Fat-soluble chemical found in vegetable oils. Deficiency of tocopherol leads to multiple adverse effects on health. In rats, vitamin E deficiency has been shown to cause sterility.

tongue

In tetrapod vertebrates, a muscular organ usually attached to the floor of the mouth. It has a thick root attached to a U-shaped bone (hyoid), and is covered with a mucous membrane containing nerves and taste buds. It is the main organ of taste. The tongue directs food to the teeth and into the throat for chewing and swallowing. In humans, it is crucial for speech; in other animals, for lapping up water and for grooming, among other functions. In some animals, such as frogs, it can be flipped forward to catch insects; in others, such as anteaters, it serves to reach for food found in deep holes.

tonsils

In higher vertebrates, masses of lymphoid tissue situated at the back of the mouth and throat (palatine tonsils), and on the rear surface of the tongue (lingual tonsils). The tonsils contain many lymphocytes and are part of the body's defense system against infection.

tooth

In vertebrates, one of a set of hard, bonelike structures in the mouth, used for biting and chewing food, and in defense and aggression. In humans, the first set (twenty milk teeth) appear from age six months to two and a half years. The permanent dentition replaces these from the sixth year onward, the wisdom teeth (third molars) sometimes not appearing until the age of twenty-five or thirty. Adults have thirty-two teeth: two incisors, one canine (eye tooth), two premolars, and three molars on each side of each jaw. Each tooth consists of an enamel coat (hardened calcium deposits), dentine (a thick, bonelike layer), and an inner pulp cavity, housing nerves and blood vessels. Mammalian teeth have roots surrounded by cementum, which fuses them into their sockets in the jawbones. The neck of the tooth is covered by the gum, while the enamel-covered crown protrudes above the gum line.

touch

Sensation produced by specialized nerve endings in the skin. Some respond to light pressure, others to heavy pressure. Temperature detection may also contribute to the overall sensation of touch. Many animals, such as nocturnal ones, rely on touch more than humans do. Some have specialized organs of touch that project from the body, such as whiskers or antennae.

toxin

Any poison produced by another living organism (usually a bacterium) that can damage the living body. In vertebrates, toxins are broken down by enzyme action, mainly in the liver.

trace element

Chemical element necessary in minute quantities for the health of a plant or animal. For example, magnesium, which occurs in chlorophyll, is essential to photosynthesis, and iodine is needed by the thyroid gland of mammals for making hormones that control growth and body chemistry.

tracer

In science, a small quantity of a radioactive isotope (form of an element) used to follow the path of a chemical reaction or a physical or biological process. The location (and possibly concentration) of the tracer is usually detected by using a Geiger-Müller counter.

trachea

Tube that forms an airway in air-breathing animals. In land-living vertebrates, including humans, it is also known as the windpipe and runs from the larynx to the upper part of the chest. Its diameter is about 1.5 cm/0.6 in and its length 10 cm/4 in. It is strong and flexible, and reinforced by rings of cartilage. In the upper chest, the trachea branches into two tubes: the left and right bronchi, which enter the lungs. Insects have a branching network of tubes called tracheae, which conduct air from holes (spiracles) in the body surface to all the body tissues. The finest branches of the tracheae are called tracheoles.

tracheid

Cell found in the water-conducting tissue (xylem) of many plants, including gymnosperms (conifers) and pteridophytes (ferns). It is long and thin with pointed ends. The cell walls are thickened by lignin, except for numerous small rounded areas, or pits, through which water and dissolved minerals pass from one cell to another. Once mature, the cell itself dies and only its walls remain.

transcription

In living cells, the process by which the information for the synthesis of a protein is transferred from the DNA strand on which it is carried to the messenger RNA strand involved in the actual synthesis.

transfusion

Intravenous delivery of blood or blood products (plasma, red cells) into a patient's circulation to make up for deficiencies due to disease, injury, or surgical intervention. Cross-matching is carried out to ensure the patient receives the right blood group. Because of worries about blood-borne disease, there is a growing interest in autologous transfusion with units of the patient's own blood "donated" over the weeks before an operation.

translation

In living cells, the process by which proteins are synthesized. During translation, the information coded as a sequence of nucleotides in messenger RNA is transformed into a sequence of amino acids in a peptide chain. The process involves the "translation" of the genetic code.

translocation
In genetics, the exchange of genetic material between chromosomes. It is responsible for congenital abnormalities, such as Down's syndrome.

transpiration
The loss of water from a plant by evaporation. Most water is lost from the leaves through pores known as stomata, whose primary function is to allow gas exchange between the plant's internal tissues and the atmosphere. Transpiration from the leaf surfaces causes a continuous upward flow of water from the roots via the xylem, which is known as the transpiration stream.

tree
Perennial plant with a woody stem, usually a single stem (trunk), made up of wood and protected by an outer layer of bark. It absorbs water through a root system. There is no clear dividing line between shrubs and trees, but sometimes a minimum achievable height of 6 m/20 ft is used to define a tree.

tree rings
Rings visible in the wood of a cut tree.

tree-ring dating
See **dendrochronology.**

tricarboxylic acid cycle
See **Krebs cycle.**

tricuspid valve
Flap of tissue situated on the right side of the heart between the atrium and the ventricle. It prevents blood flowing backward when the ventricle contracts.

triglyceride
Chemical name for fat comprising three fatty acids reacted with a glycerol.

triticale
Cereal crop of recent origin that is a cross between wheat *Triticum* and rye *Secale.* It can produce heavy yields of high-protein grain, principally for use as animal feed.

tropism, or tropic movement
The directional growth of a plant, or part of a plant, in response to an external stimulus such as gravity or light. If the movement is directed toward the stimulus it is described as positive; if away from it, it is negative. Geotropism for example, the response of plants to gravity, causes the root (positively geotropic) to grow downward, and the stem (negatively geotropic) to grow upward.

trypsin
An enzyme in the vertebrate gut responsible for the digestion of protein molecules. It is secreted by the pancreas but in an inactive form known as trypsinogen. Activation into working trypsin occurs only in the small intestine, owing to the action of another enzyme enterokinase, secreted by the wall of the duodenum. Unlike the digestive enzyme pepsin, found in the stomach, trypsin does not require an acid environment.

tuber
Swollen region of an underground stem or root, usually modified for storing food. The potato is a stem tuber, as shown by the presence of terminal and lateral buds, the "eyes" of the potato. Root tubers, for example dahlias, developed from adventitious roots (growing from the stem, not from other roots) lack these. Both types of tuber can give rise to new individuals and so provide a means of vegetative reproduction.

tumor
Overproduction of cells in a specific area of the body, often leading to a swelling or lump. Tumors are classified as benign or malignant. Benign tumors grow more slowly, do not invade surrounding tissues, do not spread to other parts of the body, and do not usually recur after removal. However, benign tumors can be dangerous in areas such as the brain. The most familiar types of benign tumor are warts on the skin. In some cases, there is no sharp dividing line between benign and malignant tumors.

turgor
The rigid condition of a plant caused by the fluid contents of a plant cell exerting a mechanical pressure against the cell wall. Turgor supports plants that do not have woody stems. Plants lacking in turgor visibly wilt. The process of osmosis plays an important part in maintaining the turgidity of plant cells.

twin
One of two young produced from a single pregnancy. Human twins may be genetically identical (monozygotic), having been formed from a single fertilized egg that splits into two cells, both of which became implanted. Nonidentical (fraternal or dizygotic) twins are formed when two eggs are fertilized at the same time.

2-hydroxypropanoic acid
See **lactic acid.**

ulna
One of the two bones found in the lower limb of the tetrapod (four-limbed) vertebrate. It articulates with the shorter radius and humerus (upper arm bone) at one end and with the radius and wrist bones at the other.

ultrafiltration

Process by which substances in solution are separated on the basis of their molecular size. A solution is forced through a membrane with pores large enough to permit the passage of small solute molecules but not large ones. Ultrafiltration is a vital mechanism in the vertebrate kidney: the cell membranes lining the Bowman's capsule act as semipermeable membranes, allowing water and substances of low molecular weight such as urea and salts to pass through into the urinary tubules but preventing the larger proteins from being lost from the blood.

umbilical cord

Connection between the embryo and the placenta of placental mammals. It has one vein and two arteries, transporting oxygen and nutrients to the developing young, and removing waste products. At birth, the connection between the young and the placenta is no longer necessary. The umbilical cord drops off or is severed, leaving a scar called the navel.

umbilicus

See **navel.**

unicellular organism

Animal or plant consisting of a single cell. Most are invisible without a microscope but a few, such as the giant ameba, may be visible to the naked eye. The main groups of unicellular organisms are bacteria, protozoa, unicellular algae, and unicellular fungi or yeasts. Some become disease-causing agents, pathogens.

urea

Waste product ($CO(NH_2)_2$) formed in the mammalian liver when nitrogen compounds are broken down. It is filtered from the blood by the kidneys, and stored in the bladder as urine prior to release. When purified, it is a white, crystalline solid. In industry it is used to make urea-formaldehyde plastics (or resins), pharmaceuticals, and fertilizers.

ureter

Tube connecting the kidney to the bladder. Its wall contains fibers of smooth muscle whose contractions aid the movement of urine out of the kidney.

urethra

In mammals, a tube connecting the bladder to the exterior. It carries urine and, in males, semen.

uric acid

Nitrogen-containing waste substance ($C_5H_4N_4O_3$), formed from the breakdown of food and body protein. It is only slightly soluble in water. Uric acid is the normal means by which most land animals that develop in a shell (birds, reptiles, insects, and land gastropods) deposit their waste products. The young are unable to get rid of their excretory products while in the shell and therefore store them in this insoluble form.

urinary system
System of organs that removes nitrogenous waste products and excess water from the bodies of animals. In vertebrates, it consists of a pair of kidneys, which produce urine; ureters, which drain the kidneys; and (in bony fishes, amphibians, some reptiles, and mammals) a bladder that stores the urine before its discharge. In mammals, the urine is expelled through the urethra; in other vertebrates, the urine drains into a common excretory chamber called a cloaca, and the urine is not discharged separately.

urine
Amber-colored fluid filtered out by the kidneys from the blood. It contains excess water, salts, proteins, waste products in the form of urea, a pigment, and some acid.

uterus
Hollow muscular organ of female mammals, located between the bladder and rectum, and connected to the Fallopian tubes above and the vagina below. The embryo develops within the uterus, and in placental mammals is attached to it after implantation via the placenta and umbilical cord. The lining of the uterus changes during the menstrual cycle. In humans and other higher primates, it is a single structure, but in other mammals it is paired.

vaccine
Any preparation of modified pathogens (viruses or bacteria) that is introduced into the body, usually either orally or by a hypodermic syringe, to induce the specific antibody reaction that produces immunity against a particular disease.

vacuole
A fluid-filled, membrane-bound cavity inside a cell. It may be a reservoir for fluids that the cell will secrete to the outside, or may be filled with excretory products or essential nutrients that the cell needs to store. Plant cells usually have a large central vacuole containing sap (sugar and salts in solution) which serves both as a store of food and as a key factor in maintaining turgor. In amebas (single-celled animals), vacuoles are the sites of digestion of engulfed food particles.

vagina
The lower part of the reproductive tract in female mammals, linking the uterus to the exterior. It admits the penis during sexual intercourse, and is the birth canal down which the baby passes during delivery.

valve

In animals, a structure for controlling the direction of the blood flow. In humans and other vertebrates, the contractions of the beating heart cause the correct blood flow into the arteries because a series of valves prevents back flow. Diseased valves, detected as "heart murmurs," have decreased efficiency. The tendency for low-pressure venous blood to collect at the base of limbs under the influence of gravity is counteracted by a series of small valves within the veins. It was the existence of these valves that prompted the seventeenth-century physician William Harvey to suggest that the blood circulated around the body.

variation

A difference between individuals of the same species, found in any sexually reproducing population. Variations may be almost unnoticeable in some cases, obvious in others, and can concern many aspects of the organism. Typically, variations in size, behavior, biochemistry, or coloring may be found. The cause of the variation is genetic (that is, inherited), environmental, or more usually a combination of the two. The origins of variation can be traced to the recombination of the genetic material during the formation of the gametes, and, more rarely, to mutation.

variegation

Description of plant leaves or stems that exhibit patches of different colors. The term is usually applied to plants that show white, cream, or yellow on their leaves, caused by areas of tissue that lack the green pigment chlorophyll. Variegated plants are bred for their decorative value, but they are often considerably weaker than the normal, uniformly green plant. Many will not breed true and require vegetative reproduction.

vascular bundle

In botany, strand of primary conducting tissue (a "vein") in vascular plants, consisting mainly of water-conducting tissues, metaxylem and protoxylem, which together make up the primary xylem, and nutrient-conducting tissue, phloem. It extends from the roots to the stems and leaves. Typically the phloem is situated nearest to the epidermis and the xylem toward the center of the bundle. In plants exhibiting secondary growth, the xylem and phloem are separated by a thin layer of vascular cambium, which gives rise to new conducting tissues.

vascular plant

Plant containing vascular bundles. Pteridophytes (ferns, horsetails, and club mosses), gymnosperms (conifers and cycads), and angiosperms (flowering plants) are all vascular plants.

vas deferens

In male vertebrates, a tube conducting sperm from the testis to the urethra. The sperm is carried in a fluid secreted by various glands, and can be transported very rapidly when the smooth muscle in the wall of the vas deferens undergoes rhythmic contraction, as in sexual intercourse.

vegetative reproduction
Type of asexual reproduction in plants that relies not on spores, but on multicellular structures formed by the parent plant. Some of the main types are stolons and runners, gemmae, bulbils, sucker shoots produced from roots (such as in the creeping thistle *Cirsium arvense*), tubers, bulbs, corms, and rhizomes. Vegetative reproduction has long been exploited in horticulture and agriculture, with various methods employed to multiply stocks of plants.

vein
In animals with a circulatory system, any vessel that carries blood from the body to the heart. Veins contain valves that prevent the blood from running back when moving against gravity. They carry blood at low pressure, so their walls are thinner than those of arteries. They always carry deoxygenated blood, with the exception of the pulmonary vein, leading from the lungs to the heart in birds and mammals, which carries newly oxygenated blood.

vena cava
Either of the two great veins of the trunk, returning deoxygenated blood to the right atrium of the heart. The superior vena cava, beginning where the arches of the two innominate veins join high in the chest, receives blood from the head, neck, chest, and arms; the inferior vena cava, arising from the junction of the right and left common iliac veins, receives blood from all parts of the body below the diaphragm.

ventral surface
The front of an animal. In vertebrates, the side furthest from the backbone; in invertebrates, the side closest to the ground. The positioning of the main nerve pathways on the ventral side is a characteristic of invertebrates.

ventricle
In zoology, either of the two lower chambers of the heart that force blood to circulate by contraction of their muscular walls. The term also refers to any of four cavities within the brain in which cerebrospinal fluid is produced.

vernalization
The stimulation of flowering by exposure to cold. Certain plants will not flower unless subjected to low temperatures during their development. For example, winter wheat will flower in summer only if planted in the previous autumn. However, by placing partially germinated seeds in low temperatures for several days, the cold requirement can be supplied artificially, allowing the wheat to be sown in the spring.

vertebra

In vertebrates, an irregularly shaped bone that forms part of the vertebral column. Children have thirty-three vertebrae, five of which fuse in adults to form the sacrum and four to form the coccyx. There are seven cervical vertebrae in the neck, twelve thoracic vertebrae in the thorax with the ribs attached, and five lumbar vertebrae in the lower back.

vertebral column

The backbone, giving support to an animal and protecting its spinal cord. It is made up of a series of bones or vertebrae running from the skull to the tail, with a central canal containing the nerve fibers of the spinal cord. In tetrapods the vertebrae show some specialization with the shape of the bones varying according to position. In the chest region the upper or thoracic vertebrae are shaped to form connections to the ribs. The backbone is only slightly flexible to give adequate rigidity to the animal structure.

vertebrate

Any animal with a backbone. The forty-one thousand species of vertebrates include mammals, birds, reptiles, amphibians, and fishes. They include most of the larger animals, but in terms of numbers of species are only a tiny proportion of the world's animals. The zoological taxonomic group Vertebrata is a subgroup of the phylum Chordata.

vestigial organ

An organ that remains in diminished form after it has ceased to have any significant function in the adult organism. In humans, the appendix is vestigial, having once had a digestive function in our ancestors.

villus

Plural *villi*, small fingerlike projection extending into the interior of the small intestine and increasing the absorptive area of the intestinal wall. Digested nutrients, including sugars and amino acids, pass into the villi and are carried away by the circulating blood.

virus

Infectious particle consisting of a core of nucleic acid (DNA or RNA) enclosed in a protein shell. Viruses are acellular and able to function and reproduce only if they can invade a living cell to use the cell's system to replicate themselves. In the process they may disrupt or alter the host cell's own DNA. The healthy human body reacts by producing an antiviral protein, interferon, which prevents the infection spreading to adjacent cells.

vitamin

Any of various chemically unrelated organic compounds that are necessary in small quantities for the normal functioning of the human body. Many act as coenzymes, small molecules that enable enzymes to function effectively. Vitamins must be supplied by the diet because the body cannot make them. They are normally present in adequate amounts in a balanced diet. Deficiency of a vitamin may lead to a metabolic disorder ("deficiency disease"), which can be remedied by sufficient intake of the vitamin. They are generally classified as water-soluble (B and C) or fat-soluble (A, D, E, and K).

vitamin A
See **retinol.**

vitamin B$_1$
See **thiamine.**

vitamin B$_2$
See **riboflavin.**

vitamin B$_6$
See **pyridoxine.**

vitamin B$_{12}$
See **cyanocobalamin.**

vitamin C
See **ascorbic acid.**

vitamin D
See **cholecalciferol.**

vitamin E
See **tocopherol.**

vitamin H
See **biotin.**

vitamin K
See **phytomenadione.**

vitreous humor
Transparent jellylike substance behind the lens of the vertebrate eye. It gives rigidity to the spherical form of the eye and allows light to pass through to the retina.

viviparous
In animals, a method of reproduction in which the embryo develops inside the body of the female from which it gains nourishment (in contrast to oviparous

and ovoviviparous). Vivipary is best developed in placental mammals, but also occurs in some arthropods, fishes, amphibians, and reptiles that have placentalike structures. In plants, it is the formation of young plantlets or bulbils instead of flowers. The term also describes seeds that germinate prematurely, before falling from the parent plant.

vivisection
Literally, cutting into a living animal. Used originally to mean experimental surgery or dissection practiced on a live subject, the term is often used by antivivisection campaigners to include any experiment on animals, surgical or otherwise.

vocal cords
The paired folds, ridges, or cords of tissue within a mammal's larynx, and a bird's syrinx. Air constricted between the folds or membranes makes them vibrate, producing sounds. Muscles in the larynx change the pitch of the sounds produced, by adjusting the tension of the vocal cords.

wall pressure
In plants, the mechanical pressure exerted by the cell contents against the cell wall. The rigidity (turgor) of a plant often depends on the level of wall pressure found in the cells of the stem. Wall pressure falls if the plant cell loses water.

warning coloration
An alternative term for aposematic coloration.

Washington Convention
Alternative name for CITES, the international agreement that regulates trade in endangered species.

water
Is a chemical compound of hydrogen and oxygen elements, H_2O. It can exist as a solid (ice), liquid (water), or gas (water vapor). Water is the most common element on Earth and vital to all living organisms. It covers 70 percent of the earth's surface, and provides a habitat for large numbers of aquatic organisms. It is the largest constituent of all living organisms—the human body consists of about 65 percent water. Pure water is a colorless, odorless, tasteless liquid which freezes at 0°C/32°F, and boils at 100°C/212°F. Natural water in the environment is never pure and always contains a variety of dissolved substances. Some 97 percent of the earth's water is in the oceans; a further 2 percent is in the form of snow or ice, leaving only 1 percent available as fresh water for plants and animals. The recycling and circulation of water through the biosphere is termed the water cycle, or "hydrological cycle"; regulation of the water balance in organisms is termed osmoregulation.

water-borne disease
Disease associated with poor water supply. In the Third World four-fifths of all illness is caused by water-borne diseases, with diarrhea being the leading cause of childhood death. Malaria, carried by mosquitoes dependent on stagnant water for breeding, affects four hundred million people every year and kills five million. Polluted water is also a problem in industrialized nations, where industrial dumping of chemical, hazardous, and radioactive wastes causes a range of diseases from headache to cancer.

water pollution
Any addition to fresh or sea water that disrupts biological processes or causes a health hazard. Common pollutants include nitrates and pesticides. A huge range of industrial contaminants, such as chemical by-products and residues created in the manufacture of various goods, also enter water—legally, accidentally, and through illegal dumping.

white blood cell, or leukocyte
One of a number of different cells that play a part in the body's defenses and give immunity against disease. Some (phagocytes and macrophages) engulf invading microorganisms, others kill infected cells, while lymphocytes produce more specific immune responses. White blood cells are colorless, with clear or granulated cytoplasm, and are capable of independent ameboid movement. They occur in the blood, lymph, and elsewhere in the body's tissues.

wild type
In genetics, the naturally occurring gene for a particular character that is typical of most individuals of a given species, as distinct from new genes that arise by mutation.

wilting
The loss of rigidity (turgor) in plants, caused by a decreasing wall pressure within the cells making up the supportive tissues. Wilting is most obvious in plants that have little or no wood.

wing
The modified forelimb of birds and bats, or the membranous outgrowths of the exoskeleton of insects, which give the power of flight. Birds and bats have two wings. Bird wings have feathers attached to the fused digits ("fingers") and forearm bones, while bat wings consist of skin stretched between the digits. Most insects have four wings, which are strengthened by wing veins.

womb
Common name for the uterus.

wood
The hard tissue beneath the bark of many perennial plants; it is composed of water-conducting cells, or secondary xylem, and gains its hardness and strength from deposits of lignin. Hardwoods, such as oak, and softwoods, such as pine, have commercial value as structural material and for furniture.

worm
Any of various elongated limbless invertebrates belonging to several phyla. Worms include the flatworms, such as flukes and tapeworms; the roundworms or nematodes, such as the eelworm and the hookworm; the marine ribbon worms or nemerteans; and the segmented worms or annelids.

X chromosome
Larger of the two sex chromosomes, the smaller being the Y chromosome. These two chromosomes are involved in sex determination. Females have two X chromosomes, males have an X and a Y. Genes carried on the X chromosome produce the phenomenon of sex linkage.

xerophyte
Plant adapted to live in dry conditions. Common adaptations to reduce the rate of transpiration include a reduction of leaf size, sometimes to spines or scales; a dense covering of hairs over the leaf to trap a layer of moist air (as in edelweiss); water storage cells; sunken stomata; and permanently rolled leaves or leaves that roll up in dry weather (as in marram grass). Many desert cacti are xerophytes.

xylem
Tissue found in vascular plants, whose main function is to conduct water and dissolved mineral nutrients from the roots to other parts of the plant. Xylem is composed of a number of different types of cell, and may include long, thin, usually dead cells known as tracheids; fibers (schlerenchyma); thin-walled parenchyma cells; and conducting vessels.

Y chromosome
Smaller of the two sex chromosomes. In male mammals it occurs paired with the other type of sex chromosome (X), which carries far more genes. The Y chromosome is the smallest of all the mammalian chromosomes and is considered to be largely inert (that is, without direct effect on the physical body). There are only twenty genes discovered so far on the human Y chromosome, much fewer than on all other human chromosomes.

yeast
One of various single-celled fungi that form masses of tiny round or oval cells by budding. When placed in a sugar solution the cells multiply and convert the sugar into alcohol and carbon dioxide. Yeasts are used as fermenting agents in baking, brewing, and the making of wine and spirits. Brewer's yeast *(S. cerevisiae)*

is a rich source of vitamin B (especially genus *Saccharomyces* and other related genera).

yeast artificial chromosome

YAC, fragment of DNA from the human genome inserted into a yeast cell. The yeast replicates the fragment along with its own DNA. In this way the fragments are copied to be preserved in a gene library. YACs are characteristically between 250,000 and 1 million base pairs in length. A cosmid works in the same way.

yolk

Store of food, mostly in the form of fats and proteins, found in the eggs of many animals. It provides nourishment for the growing embryo.

yolk sac

Sac containing the yolk in the egg of most vertebrates. The term is also used for the membranous sac formed below the developing mammalian embryo and connected with the umbilical cord.

zoidogamy

Type of plant reproduction in which male gametes (antherozoids) swim in a film of water to the female gametes. Zoidogamy is found in algae, bryophytes, pteridophytes, and some gymnosperms (others use siphonogamy).

zoology

Branch of biology concerned with the study of animals. It includes any aspect of the study of animal form and function—description of present-day animals, the study of evolution of animal forms, anatomy, physiology, embryology, behavior, and geographical distribution.

zygote

Ovum (egg) after fertilization but before it undergoes cleavage to begin embryonic development.

Nobel Prize for Physiology or Medicine

Year	Winner(s)	Awarded for
1901	Emil von Behring (Germany)	discovery that the body produces antitoxins, and development of serum therapy for diseases such as diphtheria
1902	Ronald Ross (UK)	work on the role of the *Anopheles* mosquito in transmitting malaria
1903	Niels Finsen (Denmark)	discovery of the use of ultraviolet light to treat skin diseases
1904	Ivan Pavlov (Russia)	discovery of the physiology of digestion
1905	Robert Koch (Germany)	investigations and discoveries in relation to tuberculosis
1906	Camillo Golgi (Italy) and Santiago Ramón y Cajal (Spain)	discovery of the fine structure of the nervous system
1907	Charles Laveran (France)	discovery that certain protozoa can cause disease
1908	Ilya Mechnikov (Russia) and Paul Ehrlich (Germany)	work on immunity
1909	Emil Kocher (Switzerland)	work on the physiology, pathology, and surgery of the thyroid gland
1910	Albrecht Kossel (Germany)	study of cell proteins and nucleic acids
1911	Allvar Gullstrand (Sweden)	work on the refraction of light through the different components of the eye

Nationality given is the citizenship of recipient at the time award was made.

Year	Winner(s)	Awarded for
1912	Alexis Carrel (France)	work on the techniques for connecting severed blood vessels and transplanting organs
1913	Charles Richet (France)	work on allergic responses
1914	Robert Bárány (Austria-Hungary)	work on the physiology and pathology of the equilibrium organs of the inner ear
1915	no award	
1916	no award	
1917	no award	
1918	no award	
1919	Jules Bordet (Belgium)	work on immunity
1920	August Krogh (Denmark)	discovery of the mechanism regulating the dilation and constriction of blood capillaries
1921	no award	
1922	Archibald Hill (UK)	work in the production of heat in contracting muscle
	Otto Meyerhof (Germany)	work in the relationship between oxygen consumption and metabolism of lactic acid in muscle
1923	Frederick Banting (Canada) and John Macleod (UK)	discovery and isolation of the hormone insulin
1924	Willem Einthoven (Netherlands)	invention of the electrocardiograph
1925	no award	
1926	Johannes Fibiger (Denmark)	discovery of a parasite *Spiroptera carcinoma* that causes cancer
1927	Julius Wagner-Jauregg (Austria)	use of induced malarial fever to treat paralysis caused by mental deterioration

Year	Winner(s)	Awarded for
1928	Charles Nicolle (France)	work on the role of the body louse in transmitting typhus
1929	Christiaan Eijkman (Netherlands)	discovery of a cure for beriberi, a vitamin-deficiency disease
	Frederick Hopkins (UK)	discovery of trace substances, now known as vitamins, that stimulate growth
1930	Karl Landsteiner (United States)	discovery of human blood groups
1931	Otto Warburg (Germany)	discovery of respiratory enzymes that enable cells to process oxygen
1932	Charles Sherrington (UK) and Edgar Adrian (UK)	discovery of function of neurons (nerve cells)
1933	Thomas Morgan (United States)	work on the role of chromosomes in heredity
1934	George Whipple (United States), George Minot (United States), and William Murphy (United States)	work on treatment of pernicious anemia by increasing the amount of liver in the diet
1935	Hans Spemann (Germany)	organizer effect in embryonic development
1936	Henry Dale (UK) and Otto Loewi (Germany)	chemical transmission of nerve impulses
1937	Albert Szent-Györgyi (Hungary)	investigation of biological oxidation processes and of the action of ascorbic acid (vitamin C)
1938	Corneille Heymans (Belgium)	mechanisms regulating respiration
1939	Gerhard Domagk (Germany)	discovery of the first antibacterial sulfonamide drug
1940	no award	
1941	no award	
1942	no award	

Year	Winner(s)	Awarded for
1943	Henrik Dam (Denmark)	discovery of vitamin K
	Edward Doisy (United States)	chemical nature of vitamin K
1944	Joseph Erlanger (United States) and Herbert Gasser (United States)	transmission of impulses by nerve fibers
1945	Alexander Fleming (UK)	discovery of the bactericidal effect of penicillin
	Ernst Chain (UK) and Howard Florey (Australia)	isolation of penicillin and its development as an antibiotic drug
1946	Hermann Muller (United States)	discovery that X-ray irradiation can cause mutation
1947	Carl Cori (United States) and Gerty Cori (United States)	production and breakdown of glycogen (animal starch)
	Bernardo Houssay (Argentina)	function of the pituitary gland in sugar metabolism
1948	Paul Müller (Switzerland)	discovery of the first synthetic contact insecticide DDT
1949	Walter Hess (Switzerland)	mapping areas of the midbrain that control the activities of certain body organs
	Antonio Egas Moniz (Portugal)	therapeutic value of prefrontal leucotomy in certain psychoses
1950	Edward Kendall (United States), Tadeus Reichstein (Switzerland), and Philip Hench (United States)	structure and biological effects of hormones of the adrenal cortex
1951	Max Theiler (South Africa)	discovery of a vaccine against yellow fever
1952	Selman Waksman (United States)	discovery of streptomycin, the first antibiotic effective against tuberculosis
1953	Hans Krebs (UK)	discovery of the Krebs cycle
	Fritz Lipmann (United States)	discovery of coenzyme A, a nonprotein compound that acts in conjunction with enzymes to catalyze metabolic reactions leading up to the Krebs cycle

Year	Winner(s)	Awarded for
1954	John Enders (United States), Thomas Weller (United States), and Frederick Robbins (United States)	cultivation of the polio virus in the laboratory
1955	Hugo Theorell (Sweden)	work on the nature and action of oxidation enzymes
1956	André Cournand (United States), Werner Forssmann (West Germany), and Dickinson Richards (United States)	work on the technique for passing a catheter into the heart for diagnostic purposes
1957	Daniel Bovet (Italy)	discovery of synthetic drugs used as muscle relaxants in anesthesia
1958	George Beadle (United States) and Edward Tatum (United States)	discovery that genes regulate precise chemical effects
	Joshua Lederberg (United States)	work on genetic recombination and the organization of bacterial genetic material
1959	Severo Ochoa (United States) and Arthur Kornberg (United States)	discovery of enzymes that catalyze the formation of RNA (ribonucleic acid) and DNA (deoxyribonucleic acid)
1960	Macfarlane Burnet (Australia) and Peter Medawar (UK)	acquired immunological tolerance of transplanted tissues
1961	Georg von Békésy (United States)	investigations into the mechanism of hearing within the cochlea of the inner ear
1962	Francis Crick (UK), James Watson (United States), and Maurice Wilkins (UK)	discovery of the double-helical structure of DNA and of the significance of this structure in the replication and transfer of genetic information
1963	John Eccles (Australia), Alan Hodgkin (UK), and Andrew Huxley (UK)	ionic mechanisms involved in the communication or inhibition of impulses across neuron (nerve cell) membranes
1964	Konrad Bloch (United States) and Feodor Lynen (West Germany)	work on the cholesterol and fatty-acid metabolism

Year	Winner(s)	Awarded for
1965	François Jacob (France), André Lwoff (France), and Jacques Monod (France)	genetic control of enzyme and virus synthesis
1966	Peyton Rous (United States)	discovery of tumor-inducing viruses
	Charles Huggins (United States)	hormonal treatment of prostatic cancer
1967	Ragnar Granit (Sweden), Haldan Hartline (United States), and George Wald (United States)	physiology and chemistry of vision
1968	Robert Holley (United States), Har Gobind Khorana (United States), and Marshall Nirenberg (United States)	interpretation of genetic code and its function in protein synthesis
1969	Max Delbrück (United States), Alfred Hershey (United States), and Salvador Luria (United States)	replication mechanism and genetic structure of viruses
1970	Bernard Katz (UK), Ulf von Euler (Sweden), and Julius Axelrod (United States)	work on the storage, release, and inactivation of neurotransmitters
1971	Earl Sutherland (United States)	discovery of cyclic AMP, a chemical messenger that plays a role in the action of many hormones
1972	Gerald Edelman (United States) and Rodney Porter (UK)	work on the chemical structure of antibodies
1973	Karl von Frisch (Austria), Konrad Lorenz (Austria), and Nikolaas Tinbergen (UK)	work in animal behavior patterns
1974	Albert Claude (United States), Christian de Duve (Belgium), and George Palade (United States)	work in structural and functional organization of the cell
1975	David Baltimore (United States), Renato Dulbecco (United States), and Howard Temin (United States)	work on interactions between tumor-inducing viruses and the genetic material of the cell
1976	Baruch Blumberg (United States) and Carleton Gajdusek (United States)	new mechanisms for the origin and transmission of infectious diseases

Year	Winner(s)	Awarded for
1977	Roger Guillemin (United States) and Andrew Schally (United States)	discovery of hormones produced by the hypothalamus region of the brain
	Rosalyn Yalow (United States)	radioimmunoassay techniques by which minute quantities of hormone may be detected
1978	Werner Arber (Switzerland), Daniel Nathans (United States), and Hamilton Smith (United States)	discovery of restriction enzymes and their application to molecular genetics
1979	Allan Cormack (United States) and Godfrey Hounsfield (UK)	development of the computed axial tomography (CAT) scan
1980	Baruj Benacerraf (United States), Jean Dausset (France), and George Snell (United States)	work on genetically determined structures on the cell surface that regulate immunological reactions
1981	Roger Sperry (United States)	functional specialization of the brain's cerebral hemispheres
	David Hubel (United States) and Torsten Wiesel (Sweden)	work on visual perception
1982	Sune Bergström (Sweden), Bengt Samuelsson (Sweden), and John Vane (UK)	discovery of prostaglandins and related biologically active substances
1983	Barbara McClintock (United States)	discovery of mobile genetic elements
1984	Niels Jerne (Denmark-UK), Georges Köhler (West Germany), and César Milstein (Argentina)	work on immunity and discovery of a technique for producing highly specific, monoclonal antibodies
1985	Michael Brown (United States) and Joseph L. Goldstein (United States)	work on the regulation of cholesterol metabolism
1986	Stanley Cohen (United States) and Rita Levi-Montalcini (United States-Italy)	discovery of factors that promote the growth of nerve and epidermal cells
1987	Susumu Tonegawa (Japan)	work on the process by which genes alter to produce a range of different antibodies

Year	Winner(s)	Awarded for
1988	James Black (UK), Gertrude Elion (United States), and George Hitchings (United States)	work on the principles governing the design of new drug treatment
1989	Michael Bishop (United States) and Harold Varmus (United States)	discovery of oncogenes, genes carried by viruses that can trigger cancerous growth in normal cells
1990	Joseph Murray (United States) and Donnall Thomas (United States)	pioneering work in organ and cell transplants
1991	Erwin Neher (Germany) and Bert Sakmann (Germany)	discovery of how gatelike structures (ion channels) regulate the flow of ions into and out of cells
1992	Edmond Fisher (United States) and Edwin Krebs (United States)	isolating and describing the action of the enzyme responsible for reversible protein phosphorylation, a major biological control mechanism
1993	Phillip Sharp (United States) and Richard Roberts (UK)	discovery of split genes (genes interrupted by nonsense segments of DNA)
1994	Alfred Gilman (United States) and Martin Rodbell (United States)	discovery of a family of proteins (G-proteins) that translate messages—in the form of hormones or other chemical signals—into action inside cells
1995	Edward Lewis (United States), Eric Wieschaus (United States), and Christiane Nüsslein-Volhard (Germany)	discovery of genes which control the early stages of the body's development
1996	Peter Doherty (Australia) and Rolf Zinkernagel (Switzerland)	discovery of how the immune system recognizes virus-infected cells
1997	Stanley Prusiner (United States)	discoveries, including the "prion" theory, that could lead to new treatments of dementia-related diseases, including Alzheimer's and Parkinson's diseases
1998	Robert F. Furchgott (United States), Louis J. Ignarro (United States), and Ferid Murad (United States)	discoveries concerning nitric acid as a signalling molecule in the cardiovascular system

Index

[Note: page numbers in *italics* refer to illustrations.]